DNA Sequencing

The Basics

Series editors: R. Beynon, T. A. Brown, and C. Howe
Series advisors: T. Hunt and C. F. Higgins

DNA Sequencing
T. A. Brown

Nucleic Acid Blotting
D. C. Darling and P. M. Brickell

Somatic Cell Hybrids
C. Abbott and S. Povey

DNA Sequencing
The Basics

T. A. Brown

Department of Biochemistry and Applied Molecular Biology, UMIST, Manchester

IRL PRESS
—at—
OXFORD UNIVERSITY PRESS
Oxford New York Tokyo

1994

Oxford University Press, Walton Street, Oxford OX2 6DP

Oxford New York
Athens Auckland Bangkok Bombay
Calcutta Cape Town Dar es Salaam Delhi
Florence Hong Kong Istanbul Karachi
Kuala Lumpur Madras Madrid Melbourne
Mexico City Nairobi Paris Singapore
Taipei Tokyo Toronto
and associated companies in
Berlin Ibadan

Oxford is a trade mark of Oxford University Press

Published in the United States by
Oxford University Press Inc., New York

A catalogue record for this book is available from the British Library

Library of Congress Cataloging in Publication Data
Brown, T. A. (Terence A.)
DNA sequencing: the basics / T. A. Brown.
Includes bibliographical references and index.
1. Nucleotide sequence. I. Title.
QP624.B76 1994 574.87'3282—dc20 93–44554
ISBN 0 19 963421 1 (Pbk)

Printed in Great Britain by
The Bath Press, Avon

Preface

DNA sequencing, the technique that launched a thousand kits, is described in every reputable molecular biology manual. Sequencing experiments are easy: carefully follow the instructions and you will have no problems. So why write another book about sequencing?

The success of a sequencing experiment is measured in the number of base pairs that you can read from the gel. As in any line of research, lack of progress in a sequencing project is most commonly due to technical shortcomings which lead to no data or data of such poor quality that reliable conclusions cannot be drawn. Think how quickly you could build up your master sequence if every gel was easy to read and gave you several hundred nucleotides! To exploit the full potential of DNA sequencing you have to ask why-questions: why do I add this reagent? why do I incubate for 15 and not 30 minutes? why do I use this enzyme and not that one? These questions tend to be ignored, not because their importance is unappreciated, but because the answers can be difficult to discover. Cloning manuals are very good at providing recipes and procedures, but they rarely explain the techniques from first principles. When they do provide a commentary they often assume you already know the answers to the really basic questions. So the reason for writing this book is to answer the basic questions about DNA sequencing, the ones that your colleagues in the lab assume you already know the answers to (though often they don't know the answers themselves).

This book draws on my own experiences in attempting to understand what I was doing when I first sequenced DNA. The list of people who helped me is too long to thank each one individually, but one person I will single out is Paul Towner, who explained a number of points that had always confused me. Thanks are also due to Paul Sims, Paul Birch, Jaleel Miyan and Rob Beynon, all of UMIST, who either provided Figures or helped me produce them. Finally, my wife Keri displayed her usual patience, offered advice and encouragement, and did not complain when I put the book first.

Manchester
January 1994

T. B.

Contents

CHAPTER 4 Chain termination sequencing: running the gel and reading the sequence

Abbreviations

A	2′-deoxyadenosine (in a DNA sequence)
ADP	adenosine 5′-diphosphate
ATP	adenosine 5′-triphosphate
bis	N,N'-methylenebisacrylamide
bp	base pair
C	2′-deoxycytidine (in a DNA sequence)
cDNA	complementary DNA
CD-ROM	compact disc, read-only memory
dATP	2′-deoxyadenosine 5′-triphosphate
dCTP	2′-deoxycytidine 5′-triphosphate
ddATP	2′,3′-dideoxyadenosine 5′-triphosphate
ddCTP	2′,3′-dideoxycytidine 5′-triphosphate
ddGTP	2′,3′-dideoxyguanosine 5′-triphosphate
ddNTP	2′,3′-dideoxynucleoside 5′-triphosphate
ddTTP	2′,3′-dideoxythymidine 5′-triphosphate
dGTP	2′-deoxyguanosine 5′-triphosphate
dITP	2′-deoxyinosine 5′-triphosphate
DMS	dimethyl sulphate
DNA	deoxyribonucleic acid
DNase	deoxyribonuclease
dNTP	2′-deoxynucleoside 5′-triphosphate
dTTP	2′-deoxythymidine 5′-triphosphate
EDTA	ethylenediaminetetra-acetic acid
e-mail	electronic mail
Exo	exonuclease
G	2′-deoxyguanosine (in a DNA sequence)
IPTG	isopropyl-thiogalactoside
kb	kilobase pair
Lac	lactose
ORF	open reading frame
PC	personal computer
PCR	polymerase chain reaction
PEG	polyethylene glycol
PNK	T4 polynucleotide kinase
RF	replicative form
RNA	ribonucleic acid
SDS	sodium dodecyl sulphate
SSB	single-stranded binding protein
T	2′-deoxythymidine (in a DNA sequence)
TdT	terminal deoxynucleotidyl transferase
TEMED	N,N,N',N'-tetramethylethylenediamine
tRNA	transfer RNA
u.v.	ultraviolet
X-gal	5-bromo-4-chloro-3-indolyl-β-D-galactopyranoside

1

The advent of DNA sequencing

1. The Dark Ages

Amazingly, there once was a time when it was not possible to sequence DNA. To a molecular biologist in the 1990s it is almost unimaginable that research could progress without DNA sequencing, but somehow the Dark Age scientists managed to get by. This is not to say that the potential benefits of DNA sequencing were not recognized in those days. Throughout the 1960s and early 1970s, several of the cleverest biologists of the time struggled to develop methods for sequencing nucleic acids, but the techniques that emerged from their labours were applicable mainly to RNA rather than DNA. The first nucleic acid for which a complete nucleotide sequence was worked out was the alanine-tRNA of yeast, published by Robert Holley's group in Cornell in 1965, followed during the next few years by more tRNAs, and other small RNA molecules such as the RNA genome of bacteriophage MS2.

Why were tRNAs the first nucleic acids to be sequenced? If we can understand the answer to this question we may see why DNA sequencing methods were not developed earlier. Transfer RNAs have two important features that made them ideal for sequencing:

- transfer RNAs are short (most are between 74 and 95 nucleotides in length)
- pure samples of individual tRNAs can be obtained (though not that easily: Holley's group spent three years purifying 1 mg of alanine-tRNA from 90 kg of yeast cells)

DNA molecules share neither of these features. The chromosomal DNA molecules present in human nuclei are between 55 000 000 bp (chromosome 21) and 250 000 000 bp (chromosome 1) in length, short by nobody's standards. Even today, assembling the complete nucleotide sequence of an entire chromosomal DNA molecule is a massive undertaking. Of course, a long DNA molecule can be broken into smaller fragments, but now the second problem—purification—raises its head. A molecule of about 500 bp is the longest that can be sequenced in a single experiment. A human chromosome 250 000 000 bp in length would be broken into half a million fragments 500 bp long. How on earth could one of these fragments be purified from all the others?

The answer today is by cloning or by the polymerase chain reaction. But back in the Dark Ages neither of these techniques had been invented. DNA molecules were too big to be sequenced intact, and if broken into smaller pieces there was no easy way of obtaining a pure sample of any single fragment. This meant that even if DNA sequencing techniques could be devised, obtaining the DNA to sequence would be difficult, if not impossible.

The advent of gene cloning changed everything: virtually overnight it became possible to obtain pure samples of defined fragments of chromosomal DNA, and suddenly the development of efficient DNA sequencing methods became of paramount importance. Biologists in the mid-1970s rose to the challenge and by 1977 two different but equally effective DNA sequencing techniques had been invented. These two techniques—the chain termination and chemical degradation methods—are now used, with only a few modifications to the original designs, in molecular biology labs throughout the world. It is difficult to imagine how we could ever have managed without them.

The aim of this book is to guide the newcomer into the wonderful world of sequencing. If you are starting out in molecular biology research then you will probably run your first sequencing experiment within the next few months. If you stay in research for any length of time then you will probably carry out tens if not hundreds of sequencing experiments during your career. Most research papers in molecular biology either present a new DNA sequence or refer to an existing one. DNA sequencing is therefore a technique you are going to have to get to grips with, so let's get started!

2. DNA sequencing in outline

DNA sequencing is a complicated business and in the next few chapters we have to deal with the complexities of such things as DNA polymerase processivities, gradient gels, and base modification chemicals. To be able to understand these topics we must first have a clear picture in our minds of what happens during a DNA sequencing experiment. So we will start in this chapter by examining chain termination and chemical degradation sequencing in outline, concentrating on the basic principles and leaving the complicated details until later.

2.1 The basic idea behind DNA sequencing

Although the two modern sequencing techniques are very different they both work in accordance with the same fundamental principle. This principle can be summarized by the statement that two single stranded DNA molecules that differ in length by just a single nucleo-

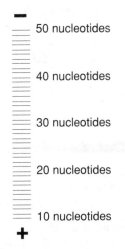

Fig 1.1

A ladder of bands produced when a family of single stranded DNA molecules representing all lengths between 10 and 50 nucleotides are separated by polyacrylamide gel electrophoresis, using the conditions described in Chapter 4. The negative electrode is at the top and the positive electrode at the bottom

◇ In Chapter 2 we will see how the single stranded template molecules are obtained. Exactly how the chain termination sequencing reactions are carried out will be described in Chapter 3.

tide can be separated into distinct bands by electrophoresis in a polyacrylamide gel. This means that a family of molecules representing all possible lengths from, say, 10 to 50 nucleotides will form a ladder of bands after polyacrylamide gel electrophoresis (*Figure 1.1*).

2.1.1 The chain termination method

How does this help us in DNA sequencing? We will examine the chain termination method to see the answer. The starting material for this procedure is a preparation of single stranded DNA molecules, all the molecules exactly the same. The first step is to attach ('anneal') a short oligonucleotide on to each single stranded molecule (*Figure 1.2A*). This oligonucleotide acts as the starting point (the 'primer') for synthesis of a new polynucleotide chain, complementary to the existing ('template') strand (*Figure 1.2B*). The chain elongation reaction is catalysed by a DNA polymerase enzyme and requires the four deoxyribonucleotide triphosphates (dATP, dCTP, dGTP, and dTTP) as substrates. Normally the DNA polymerase enzyme would be able to make a new DNA chain several thousand nucleotides in length, but in a sequencing experiment the length of the chain is limited because, as well as the four dNTPs, you also add a modified nucleotide into the reaction. This modified nucleotide is called a dideoxynucleotide (e.g. ddATP). It can be incorporated into the growing polynucleotide chain just as efficiently as the normal nucleotide, but it blocks further chain elongation (*Figure 1.2C*). This is because the dideoxynucleotide lacks the hydroxyl group at the 3' position on the sugar component (*Figure 1.3*). This group is needed for the next nucleotide to be attached; chain termination therefore occurs immediately a dideoxynucleotide is incorporated into the growing DNA molecule.

A. Anneal the primer to the template

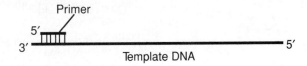

B. Strand synthesis

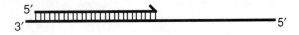

C. Strand termination

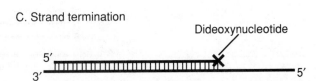

Fig 1.2

The strand synthesis reaction that forms the basis to chain termination sequencing

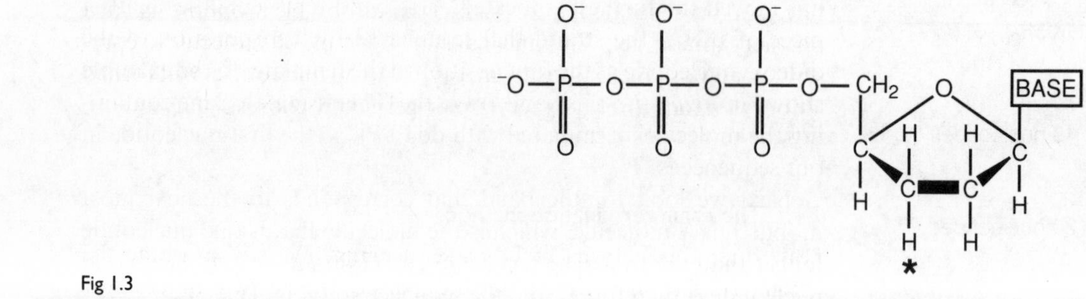

Fig 1.3

2′,3′-dideoxynucleotide 5′-triphosphate
(ddNTP)

* Position where the -OH of a
dNTP is replaced by -H

If ddATP is added to the reaction mixture, then chain termination occurs at positions opposite thymidines in the template (*Figure 1.4A*). But termination does not always occur at the first T as normal dATP is also present and may be incorporated instead of the dideoxynucleotide. The ratio of dATP to ddATP is such that an individual chain may in fact be polymerized for a considerable distance before the ddATP is added. On the other hand, the ddATP may slip in at the first thymidine that is encountered. The result is therefore a family of new chains, all of different lengths, but each ending in ddATP (*Figure 1.4B*).

The strand synthesis reaction is carried out four times in parallel. As well as the reaction with ddATP, there is one with ddCTP, one with ddGTP, and one with ddTTP. You therefore end up with four distinct families of newly synthesized polynucleotide chains, one family containing strands ending in ddATP, one with strands ending in ddCTP, and so on.

Now the gel electrophoresis step comes in. Each family of molecules is loaded into a different lane of a polyacrylamide gel and electrophoresed under the conditions that separate DNA molecules one nucleotide different in length. The sequence can then be read directly from the positions of the bands in the gel. First the band that

A. Termination with ddATP . . .

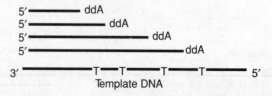

B. . . . leads to a family of chains

Fig 1.4

Chain termination

◇ You will find out everything you need to know about reading a sequence in Chapter 4.

has moved the furthest is located. This band represents the smallest piece of DNA, the strand that terminated by incorporation of the dideoxynucleotide at the first position in the template. In the example shown in *Figure 1.5* this band lies in the 'T' lane (i.e. the lane containing the molecules terminated with ddTTP), so the first nucleotide in our sequence is 'T'.

Next we look for the band that corresponds to the next most mobile DNA molecule, which is the molecule that is one nucleotide longer than the first. The lane is noted, 'C' in *Figure 1.5*; the second nucleotide is therefore C and the sequence so far is 'TC'.

Continuing up through the gel we find that the next band also lies in the C lane (sequence TCC), then we move to the A lane (TCCA), then T (TCCAT), followed by two Gs, a C, and three more Gs (TCCATGGCGGG). What could be easier?

2.1.2 The chemical degradation method

The basic difference between the chemical degradation and chain termination techniques lies in the way in which the A, C, G, and T families of molecules are generated. In a chemical degradation experiment these families are produced not by synthesizing new chains, but by breaking down the starting molecules. This is achieved with chemicals that cleave specifically at a particular nucleotide. For example, the G family is generated by treating the DNA molecules with a combination of dimethyl sulphate and piperidine, which results in the polynucleotides being cut at guanosine residues. The reaction is carried out under limiting conditions, so that on average just one cut is made in each polynucleotide. Once the A, C, G, and T families of cleaved molecules have been obtained they are separated by gel electrophoresis, and the sequence read, in a manner very similar to chain termination sequencing.

◇ Don't worry if it is not clear to you at this stage how strand cleavage results in a readable banding pattern after electrophoresis. We will go through the process in more detail in Chapter 5.

Fig 1.5

Reading the DNA sequence from the autoradiograph produced by a chain termination experiment

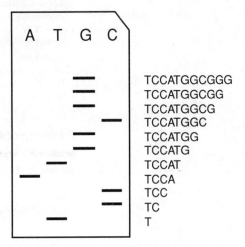

3. Why this book is mainly about chain termination sequencing

The two DNA sequencing methods were invented on different sides of the Atlantic, the chemical degradation method by Walter Gilbert and Allan Maxam of Harvard, and the chain termination method by Frederick Sanger and Andrew Coulson of the MRC Laboratory for Molecular Biology at Cambridge. Both methods were first published in 1977 (Maxam and Gilbert 1977; Sanger *et al.* 1977) and by about 1980 the techniques were commonplace in the better molecular biology labs. The chemical degradation method was responsible for the complete sequences of the SV40 virus genome (5243 bp) and the cloning vector pBR322 (4363 bp) both published in 1978, and the chain termination method was used with the human mitochondrial genome (16.5 kb) and the DNA of bacteriophage λ (49 kb), published in 1981 and 1982 respectively.

At one time the chemical degradation method was more popular but gradually during the 1980s most labs switched to chain termination sequencing. The main reason for this was the introduction of innovations to the original chain termination technique, which meant that longer sequences could be obtained. Today, chain termination sequencing is chosen for most 'routine' applications, and it is probably the technique that you will yourself make use of. The bulk of this book is therefore about the chain termination method, with chemical degradation sequencing being put on hold until Chapter 5.

◇ The length of sequence that can be obtained in a single experiment depends on a number of factors, not least the skill of the sequencer. In general, it is possible to be confident of about 500 bp of sequence from a single chain termination experiment, compared with about 250 bp for a chemical degradation experiment. The chemical degradation method is, however, better for certain applications (see Chapter 5, Section 5).

Further reading

Ausubel, F.M. *et al.* (ed.), (first published in 1987 but updated every 3 months). *Current protocols in molecular biology*. Greene Publishing Associates and John Wiley and Sons, New York—of the many advanced cloning manuals on the market, this is probably the best.

Brown, T.A. (1990). *Gene cloning: an introduction*, (2nd edn). Chapman and Hall, London—background details on gene cloning and techniques such as DNA sequencing.

Brown, T.A. (ed.) (1991). *Essential molecular biology: A practical approach*, Vols I and II. IRL Press at Oxford University Press—a beginner's guide to molecular biology techniques.

Howe, C.J. and Ward, E.S. (ed.) (1989). *Nucleic acids sequencing: A practical approach*. IRL Press at Oxford University Press—detailed protocols for all aspects of DNA sequencing.

References

Maxam, A.M. and Gilbert, W. (1977). A new method for sequencing DNA. *Proceedings of the National Academy of Sciences, USA*, **74**, 560.

Sanger, F., Nicklen, S., and Coulson, A.R. (1977). DNA sequencing with chain-terminating inhibitors. *Proceedings of the National Academy of Sciences, USA*, **74**, 5463.

2

Chain termination sequencing: preparing the single stranded DNA template

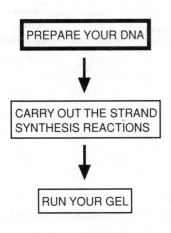

PREPARE YOUR DNA

↓

CARRY OUT THE STRAND SYNTHESIS REACTIONS

↓

RUN YOUR GEL

◇ It is possible to sequence double stranded DNA (see Chapter 3, Section 2.2.2), but clearer and longer sequences are obtained with single stranded DNA as the template. You must therefore master the techniques for obtaining single stranded DNA if you wish to use chain termination sequencing to its full potential.

There are three stages to a chain termination sequencing experiment:

Stage 1: Preparing the single stranded DNA that will be the template for the strand synthesis reactions.

Stage 2: Carrying out the strand synthesis reactions.

Stage 3: Separating the chain terminated molecules by polyacrylamide gel electrophoresis, reading the gel, and working out what it all means.

In the next three chapters we will run through each of these three stages, starting with how to prepare the template DNA.

1. M13 cloning vectors for making single stranded DNA

When the chain termination method was first published it was recognized that a major difficulty with the technique would be the need for the template DNA to be single stranded. Most cloning experiments, then and now, make use either of plasmid vectors, such as pBR322 and the members of the pUC series, or vectors based on the bacteriophage λ, λEMBL3, and λgt11, for example. Plasmids and λ vectors are double stranded molecules, so genes that are cloned in them end up as double rather than single stranded DNA. A special type of cloning vector is therefore needed for DNA sequencing. Thanks to the ingenuity and hard work of a number of molecular biologists, notably Joachim Messing of Rutgers University, vectors that produce single stranded versions of cloned genes were quickly developed. These vectors are almost exclusively based on the bacteriophage

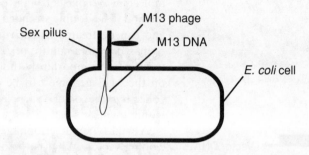

A. M13 injects its DNA into the bacterium

M13 phage

Sex pilus

M13 DNA

E. coli cell

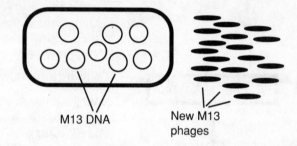

B. M13 DNA replicates and new phages are released

M13 DNA

New M13 phages

Fig 2.1

Infection of E. coli by bacteriophage M13

called M13. Before we go any further we must therefore find out about M13.

1.1 The life cycle of M13 phage

M13 is a filamentous phage with a genome 6407 nucleotides in length. Within the phage particles this genome exists as single stranded DNA (Marvin and Wachtel 1975).

M13 phage particles infect *Escherichia coli* bacteria by injecting their DNA into the cell via the sex pilus (*Figure 2.1A*), the structure that holds two bacteria together during sexual reproduction. Only 'male' bacteria (i.e. those harbouring an F plasmid) have sex pili so M13 is an example of a 'male-specific phage'. Once inside the cell the single stranded M13 DNA is converted into a double stranded molecule which replicates until over 100 copies are present (*Figure 2.1B*). This double stranded version of the phage genome is called the replicative form or RF. When the bacterium divides, each daughter receives copies of the phage RF, which continues to replicate, maintaining its overall numbers per cell. As shown in *Figure 2.1B*, new M13 particles, containing the single stranded DNA genome, are continuously assembled and released, about 1000 new phage particles being produced during each generation of the bacterium.

By now you have probably realized that M13 was a godsend for molecular biologists trying to use the chain termination sequencing method. The double stranded RF behaves very much like a plasmid and can be treated as such for experimental purposes. It is easily prepared from a culture of *E. coli* cells and can be reintroduced by transfection. Importantly though, a gene cloned into a vector based on the M13 genome can be obtained as single stranded DNA by collecting the phage particles and purifying the DNA contained within them.

◇ If you are not sure about terms such as 'transfection' then check the Glossary.

A wild-type bacteriophage is not, however, the same thing as a cloning vector. What is the difference between M13 the phage and M13 the sequencer's friend?

1.2 The M13mp series of cloning vectors

The vectors constructed by Messing and his colleagues are called the M13mp series. Each of these vectors has a different set of unique restriction sites (*Table 2.1*) that can be used for the insertion of new DNA during a cloning experiment. The restriction sites are contained within a copy of the *lacZ'* gene that has been transferred to the M13 genome.

What is the function of *lacZ'*? If you have carried out cloning experiments with plasmid vectors then you probably know that this gene provides a selectable marker that enables recombinant *E. coli* bacteria (i.e. those containing vector molecules that carry cloned DNA) to be identified. The *lacZ'* gene codes for the first 146 amino acids of the *E. coli* β-galactosidase enzyme. This is the enzyme responsible for converting lactose into glucose plus galactose. The segment coded by *lacZ'* is not by itself sufficient to catalyse the conversion, but it can complement a second peptide, containing the remainder of the β-galactosidase protein, to produce an active enzyme (Ullman and Perrin 1970). This second peptide is made by the host bacterium used in the cloning experiment (*Figure 2.2*). After transformation with a *lacZ'* vector the bacteria are plated on agar

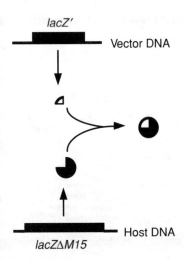

lacZ' — Vector DNA

lacZΔM15 — Host DNA

◢ Non-functional peptide coded by the vector gene

◗ Non-functional protein coded by the host gene

◔ Functional ß-galactosidase

Fig 2.2

The *lacZ'* and *lacZΔM15* genes code for complementing segments of the β-galactosidase enzyme

Table 2.1 The M13mp series of vectors

Vector	Order of restriction sites in the polylinker*
M13mp7	EcoRI—BamHI—SalI—PstI—SalI—BamHI—EcoRI
M13mp8	EcoRI—SmaI—BamHI—SalI—PstI—HindIII
M13mp9	HindIII—PstI—SalI—BamHI—SmaI—EcoRI
M13mp10	EcoRI—SstI—SmaI—BamHI—XbaI—SalI—PstI—HindIII
M13mp11	HindIII—PstI—SalI—XbaI—BamHI—SmaI—SstI—EcoRI
M13mp18	EcoRI—SstI—KpnI—SmaI—BamHI—XbaI—SalI—PstI—SphI—HindIII
M13mp19	HindIII—SphI—PstI—SalI—XbaI—BamHI—SmaI—KpnI—SstI—EcoRI

*The *SalI* sites are also cut by *AccI* and *HincII*; the *SmaI* sites are also cut by *XmaI*.

Non-recombinant cell - intact *lacZ'* gene

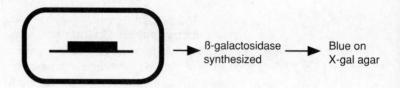

Recombinant cell - disrupted *lacZ' gene*

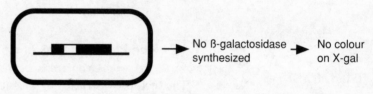

Fig 2.3

Lac selection

◇ The agar must also contain an inducer of *lacZ'*, usually IPTG—isopropyl-thiogalactoside.

◇ If you are interested in finding out more about how the M13mp vectors were constructed, then see Messing *et al.* (1981).

containing the lactose analogue X-gal (5-bromo-4-chloro-3-indolyl-β-D-galactopyranoside; Horwitz *et al.* 1964). The 'Lac selection' system then works as follows (*Figure 2.3*):

- Cells that contain a cloning vector with an intact *lacZ'* gene (i.e. no inserted DNA) are able to synthesize β-galactosidase, which converts the X-gal into bromochloroindole, a compound with a beautiful deep blue colour. These cells therefore give rise to blue colonies (with a plasmid vector) or blue plaques (with a phage vector).
- Conversely, recombinant cells, in which the *lacZ'* gene is disrupted by the inserted DNA, cannot synthesize β-galactosidase and so give rise to white colonies or clear plaques.

The *lacZ'* gene therefore provides an easy means of identifying recombinant bacteria.

When each M13mp vector was constructed, cloning sites were created within the *lacZ'* gene by inserting a synthetic oligonucleotide (a 'polylinker') consisting of an array of restriction sequences. With each vector, the polylinker is designed so that it does not disrupt the reading frame of the *lacZ'* gene, so a functional (though slightly altered) β-galactosidase enzyme is still produced by bacteria containing the vector.

If you look again at *Table 2.1* you will see that certain pairs of vectors are related in having the same order of restriction sites but in opposite directions. M13mp8 and 9 form such a pair, as do M13mp10 and 11, and M13mp18 and 19 (Yanisch-Perron *et al.* 1985). These vector pairs are particularly useful as they enable sequences to be obtained from both ends of a cloned fragment. As we will see in Chapter 3, during a chain termination experiment the

A. Shuttling a fragment from M13mp18 to M13mp19 . . .

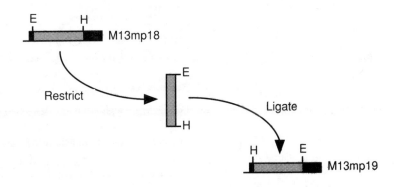

B. . . . enables an overlapping sequence to be obtained

Fig 2.4

Vector pairs enable sequences to be obtained from both ends of an inserted fragment. Abbreviations: E, *EcoRI* site; H, *HindIII* site

DNA sequence 1

DNA sequence 2

sequence of the inserted fragment is read from its 3′ end. If the fragment is more than about 500 bp in length then it will not be possible to obtain its complete DNA sequence in a single experiment. One answer is to turn the fragment around, by excising and re-inserting into the sister vector, as shown in *Figure 2.4A*. A second DNA sequencing experiment now provides another 500 bp, from the other end of the fragment, and with luck the two sequences will overlap in the middle (*Figure 2.4B*).

◇ Alternatively, if the fragment is too long to sequence in a single experiment then you may wish to subclone it—see Section 1.4.3.

1.3 Host bacteria for the M13mp vectors

Now that we understand how M13 vectors work we must find a host strain of bacteria for our cloning experiment. It is always essential that you choose a host bacterium that has genetic features compatible with the type of vector that you are using. From what you have learnt so far in this chapter you will realize that the host bacteria for an M13 vector must satisfy two key requirements:

- they must be male, so that they can be infected with M13 phage
- they must able to participate in Lac selection

There are a number of *E. coli* strains that fulfil these requirements, the most popular ones being those listed in *Table 2.2*. How do we know that these are the correct strains to use? The answer is because

Table 2.2 A few of the *E. coli* host strains suitable for M13 cloning experiments

Strain	Genotype
JM109	*recA1 endA1 gyrA96 thi hsdR17 supE44 relA1 Δ(lac-proAB)* *F'[traD36 proAB⁺ lacI^q lacZΔM15]*
JM110	*rpsL thr leu thi lacY galK galT ara tonA tsx dam dcm supE44 Δ(lac-proAB) F'[traD36 proAB⁺ lacI^q lacZΔM15]*
TG2	*supE hsdΔ5 thi Δ(srl-recA)306::Tn10(tet^r) Δ(lac-proAB) F'[traD36 proAB⁺ lacI^q lacZΔM15]*
XL1-BLUE	*supE44 hsdR17 recA1 endA1 gyrA46 thi relA1 lac F'[proAB⁺ lacI^q lacZΔM15 Tn10 (tet^r)]*

the genotype of the strain tells you exactly what characteristics the bacterium possesses, as we will see by taking *E. coli* TG2 as an example.

1.3.1 The genetic characteristics of *E. coli* TG2

The genotype of *E. coli* TG2 is written as follows:

supE hsdΔ5 thi Δ(srl-recA)306::Tn10(tet^r) Δ(lac-proAB) F'[traD36 proAB⁺ lacI^q lacZΔM15]

What does it all mean? The first thing to appreciate is that each 'word' of the genotype describes a different genetic feature of the bacterium. The TG2 genotype consists of six 'words', whose meanings are as follows:

supE

This is an 'amber suppressor' mutation. A bacterium carrying this mutation recognizes the triplet UAG (the 'amber' termination codon) as a glutamine codon, so when it encounters a UAG codon during protein synthesis it inserts a glutamine amino acid into the growing polypeptide chain. A normal bacterium would terminate protein synthesis at this point. In the past, M13 cloning vectors were designed with UAG codons present within several of the important genes in the phage genome. These cloning vectors gave rise to phage particles only in laboratory strains of bacteria, which carry the suppressor mutation, and were unable to escape from the laboratory because their genes were not translated corrected by wild bacteria, living in your intestine for example. The idea was to prevent the release of genetically engineered bacteria into the environment.

◇ Nowadays, the genetic manipulation regulations have changed and M13 cloning vectors no longer have to carry amber mutations, but many of the host strains still possess the appropriate suppressors.

hsdΔ5

This indicates that TG2 has a deletion (shown by the 'Δ') in the region of its genome containing the *hsd* genes. These genes code for modification and restriction enzymes, which normally enable it to cleave foreign (e.g. phage) DNA that invades the cell. Clearly, we do not want this to happen with our cloning vector, so in TG2 the *hsd* genes are inactivated.

thi

TG2 is mutated in the *thi* gene, involved in biosynthesis of thiamine (vitamin B_1). This is not important as far as cloning is concerned, but it does mean that the bacteria must be supplied with thiamine in their growth media. Most of the time you use rich media (such as LB, YT, or nutrient broth) which contain plenty of thiamine, so you do not need to add any more. However, if you keep stocks of TG2 bacteria on minimal agar plates then these must have thiamine added.

Δ(srl-recA)306::Tn10(tet^r)

◇ The number '306' distinguishes this deletion from the 305 (at least) other deletions that have been made in this region by various geneticists over the years.

This is a more complicated 'word', in fact two words joined together. Δ(*srl-recA*)306 is another deletion within the bacterial genome, this time involving a segment stretching from the *srl* genes (for sorbitol utilization) through to the *recA* gene. The fact that TG2 is unable to use sorbitol as an energy source is immaterial, the important thing is the deletion of the *recA* gene. The *rec* genes code for the major recombination pathways of *E. coli*. Strains that possess active *rec* genes are less suitable for cloning as they can rearrange DNA fragments that have been inserted into the vector. There are at least nine different *rec* genes but, in practice, a strain mutated in *recA* is sufficiently deficient in recombination to be able to act as a host for cloned DNA.

The second part of the genetic designation—::Tn*10*(*tet*^r)—states that transposon Tn*10*, carrying a gene for tetracycline resistance, is inserted at this point in the genome. TG2 is therefore able to grow in media containing tetracycline.

Δ(lac-proAB)

A third deletion, this one removing the lactose operon and two genes involved in proline biosynthesis. These genes are not completely missing though, as the next part of the genotype tells us . . .

F'[traD36 proAB⁺ lacI^q lacZΔM15]

This is the most complicated part of the genotype. First, F' is the designation for a strain that possesses an F' plasmid, a fertility plasmid with some additional DNA inserted. The fertility plasmid carries, among other things, genes for synthesis of the sex pili that M13 phage attach to when they infect a cell (see *Figure 2.1A*). The F' plasmid therefore confers 'maleness' on TG2.

The words within the square brackets refer to genes present on the F' plasmid. *TraD36* is a mutation in one of the standard fertility plasmid genes, preventing TG2 bacteria from participating in conjugation. During conjugation plasmids can be passed from one cell to another, which could lead to a genetically engineered plasmid being transferred to wild populations of bacteria, should a transformed bacterium escape from its Petri dish. So the *traD36* mutation, by abolishing conjugation, provides another safety feature to prevent accidents.

The next word is $proAB^+$. This indicates that the proline biosynthesis genes that were deleted from the TG2 genome are in fact present on the F′ plasmid. The bacteria are therefore able to make proline, so this amino acid does not need to be supplied in the growth medium. Note that this situation holds only if the bacteria retain the F′ plasmid. Should a bacterium lose the plasmid, as might happen if an agar culture is stored for a long time, then it reverts to $proAB^-$ and is able to grow only when proline is provided. If the F′ plasmid is lost then the TG2 cells no longer make the all important sex pili that we need for M13 infection. So a TG2 culture is usually stored on minimal medium, containing no proline, which means that only those cells that retain the F′ plasmid and the ability to be infected by M13 are able to survive.

The remaining two words—$lacI^q$ and $lacZ\Delta M15$—refer to lactose genes that were deleted from the bacterial genome and have been transferred to the F′ plasmid. $LacI^q$ designates an alteration in the gene for the lactose repressor; it is not important as far as the use of TG2 in DNA sequencing is concerned. $LacZ\Delta M15$, on the other hand, is very important. You will remember that $lacZ$ is the gene for β-galactosidase, which is used in Lac selection (see Section 1.2). The $\Delta M15$ designation means that this version of the $lacZ$ gene lacks codons 11 to 41. These codons specify a part of the protein called the 'α-peptide', the absence of which leads to a non-functional β-galactosidase enzyme. In order to make the active enzyme TG2 cells must also possess the $lacZ'$ gene, which codes for the missing peptide. The $lacZ'$ gene is carried by the M13mp cloning vector, so only TG2 cells containing a cloning vector (without inserted DNA) can make active β-galactosidase. The bacterial and vector genes therefore work together in Lac selection.

◇ The $lacI^q$ mutation is beneficial in biotechnology projects as it allows greater control over the synthesis of foreign proteins from genes fused to $lacZ'$.

1.4 Working with M13mp cloning vectors

In a typical cloning experiment with an M13mp vector you carry out the following steps (*Figure 2.5*):

- ligate the inserted DNA into the double stranded replicative form of the cloning vector
- introduce the recombinant cloning vector into the host bacteria by transfection
- plate the transformants on to agar containing X-gal and IPTG

After incubating the plates overnight at 37°C you should have a collection of blue and clear plaques. Think about the basis to Lac selection (Section 1.2) and you will realize that the clear plaques are the ones you want, as these contain phage particles that carry the cloned DNA.

You now have two choices. On the one hand, you may wish to purify double stranded RF molecules from the clear plaques. This

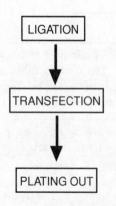

Fig 2.5

The steps in an M13 cloning experiment

◇ You may be wondering why M13 infection gives rise to plaques when the bacteria are not lysed by the phage. In fact, M13 causes an appreciable decrease in the growth rate of the infected cells, sufficient to produce a zone of clearing on a lawn of uninfected bacteria. These zones of clearing (shown below) contain infected cells plus secreted phage particles, and are visually similar to the true plaques produced by a lytic phage.

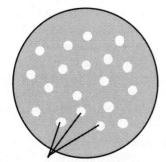

Plaques contain infected
cells plus secreted phages

will be your objective if the cloned DNA fragment is too long to be sequenced in a single experiment, as you will have to subclone the inserted DNA from double stranded RF molecules in order to obtain shorter, overlapping fragments from which to build up a master sequence. We will deal with these techniques later, in Sections 1.4.2 and 1.4.3.

First we will think about your second choice. This is where you prepare single stranded DNA from a clear plaque, so you can carry out a sequencing experiment immediately. Even if you know that the cloned DNA is too long to be sequenced in one go you probably want to obtain as much sequence as you can without delay, as even a short segment of the sequence may be enough to tell you if the cloned DNA fragment is the one you have been looking for. In any cloning project it is always good to confirm that things have progressed according to plan before expending a lot of time in fully characterizing the cloned DNA.

1.4.1 Preparation of single stranded DNA from an M13 plaque

You may already have purified DNA of one kind or another, possibly genomic DNA from animal or plant tissues, or plasmid DNA from a culture of bacteria. You will know that DNA preparation can be a messy business. You will be pleasantly surprised that it is relatively easy to obtain good quality, pure, single stranded DNA from M13 plaques. Two factors work in your favour:

- infected cells continually secrete M13 particles so it is easy to obtain enough starting material for your preparation
- M13 phage particles are made just of DNA and protein, so purification of the DNA is a relatively simple business

The sequence of steps is illustrated in *Figure 2.6*. The first thing you must do is grow a small liquid culture of infected cells and collect the phage particles that are produced. The culture does not need to be very large, just a few ml, as you need less than 1 µg of single stranded DNA for a sequencing experiment. The easiest way to inoculate the

Fig 2.6

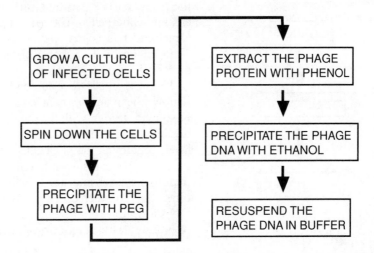

The steps involved in purification of single stranded DNA from M13 phage particles

◇ To aid aeration you should grow the culture in a tube that is at least 10 times bigger than the volume of the growth medium, and you should angle the tube on the shaker so the culture splashes about inside it as much as possible.

◇ For methods of assaying the phage titre see Bainbridge (1991).

◇ To check the size of the inserted DNA, treat 20 μl of phage supernatant with SDS, which strips away the phage protein coat leaving a crude DNA preparation that can be analysed in an agarose gel. See Messing and Bankier (1989).

◇ Make sure you use PEG 6000 or 8000. The numbers refer to the molecular weights of the PEG polymers.

culture is with a toothpick, transferring a single plaque from the agar plate to the liquid medium. Some protocols recommend inoculating into fresh medium, so the resulting culture is derived solely from the bacteria in the plaque. Other methods suggest that you add the plaque into media that has already been inoculated with uninfected bacteria, so there are new cells immediately available for the phage to infect. Whichever method you use, as soon as the plaque has been transferred the infected culture must be incubated at 37°C with vigorous aeration. The idea is to grow the culture to the highest possible cell density without allowing the cells to enter stationary phase. If things go wrong and the culture grows too far then the cells start to die, releasing bacterial RNA and DNA into the culture medium. These contaminants may co-purify with the M13 DNA, causing you endless problems later on. The high aeration prevents the cells from reaching stationary phase prematurely due to oxygen starvation. It takes practice to know exactly how long to incubate the culture but something between 5 and 8 hours is usually sufficient. You should aim for a phage titre approaching 2×10^{12} M13 particles per ml.

When your culture has reached the required stage you can begin the DNA preparation. This is usually carried out with 1.5 ml of culture, which just fits into a standard sized microfuge tube. Alternatively, if you have a large number of preparations to do at once, you can use 200 μl aliquots pipetted into the wells of a microtitre tray. The bacteria are pelleted by centrifugation and the phage particles collected by taking the supernatant. You should be careful not to resuspend any of the bacteria when you are removing the supernatant, as it is important not to contaminate the phage suspension with cellular material. Similarly, during the following steps you should do everything as carefully as possible. How pretty your sequence looks depends on the how well you deal with the DNA preparation.

Having decanted the phage suspension the next step is to obtain the phage particles themselves. This is done by adding polyethylene glycol (PEG) and salt to the supernatant. These chemicals precipitate the phage particles so they can be collected by centrifugation as a creamy white pellet. The salt used is usually sodium chloride, but sometimes this is replaced by sodium acetate, which is easier to remove later on. Carry over of salt into your 'pure' DNA is a common cause of poor sequences.

The phage pellet is resuspended in buffer, and phenol added to remove the phage protein coats. After a brief centrifugation the mixture partitions, with the phenol as the lower layer and the protein precipitated at the interface. You should remove the upper, aqueous layer, containing the phage DNA, as quickly as possible, before the protein starts to resuspend. You must be very careful not to disturb the protein layer and not to withdraw any of the phenol from the tube. If you do then centrifuge the tube again. Phenol in your 'pure' DNA is death to your sequence. Some protocols advise following the phenol step with a chloroform extraction, to remove any phenol that

◇ Always use gloves when handling phenol, and make sure they are of a suitable type (Brown 1991).

you accidentally take up or which may still remain in the aqueous layer. In fact chloroform is just as bad as phenol if it gets into your template DNA, so it is best to avoid it.

Now the DNA can be precipitated with ethanol and dissolved in a small amount of buffer. You will probably peer into the bottom of your tube after removing the ethanol, hoping to see your DNA. Don't be disappointed—if you **can** see anything then something has gone wrong, probably carry over of salt from the PEG precipitation step. If everything has progressed according to plan then you have an invisible pellet of pure single stranded DNA. Start sequencing!

◇ If you need reassurance that you have some DNA then run one-eighth of your preparation in an agarose gel and stain with ethidium bromide.

1.4.2 Preparation of double stranded RF DNA from an M13 plaque

Now we must return to the second of your choices and look at the procedures involved in preparation of the double stranded replicative form of M13. We will not spend long on this topic as the methodology is almost identical to that used to prepare plasmid DNA from transformed *E. coli* cells. You can choose any of the standard mini-prep techniques, such as the alkaline lysis method (Birnboim and Doly 1979), which provide you with a few µg of M13 RF DNA from a 3–5 ml culture. This is usually enough for the subcloning experiments needed to obtain the complete sequence of a lengthy DNA molecule. However, there is a problem with mini-preps in that the resulting DNA is sometimes of questionable purity, which leads to difficulties with the restriction and ligation reactions involved in subcloning. In the long run, it is often better to start with a more time-consuming large-scale DNA preparation, including a caesium chloride density gradient step. This gives you more DNA than you need (50–100 µg from a 50 ml culture) but the DNA is much purer than that obtained from a quick mini-prep. To inoculate the cultures from which the RF DNA will be prepared all you have to do is transfer a plaque into liquid medium, with a toothpick as described in Section 1.4.1.

◇ For details on plasmid preps, including methods for improving the purity of the DNA obtained from mini-preps, see Towner (1991).

1.4.3 Subcloning a DNA molecule in order to obtain overlapping fragments

As mentioned previously, a single chain termination experiment is unlikely to yield more than 500 bp of DNA sequence. For a piece of DNA longer than this you must sequence a series of overlapping fragments derived from the starting molecule (*Figure 2.7*). There are three different ways of obtaining overlapping fragments.

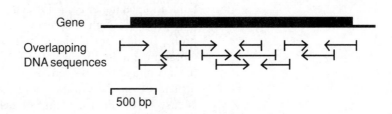

Fig 2.7

Building up a long DNA sequence

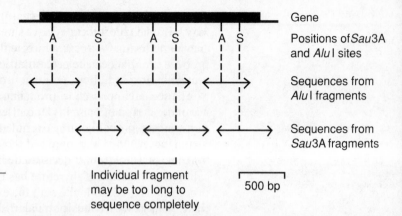

Gene

Positions of *Sau*3A and *Alu*I sites

Sequences from *Alu*I fragments

Sequences from *Sau*3A fragments

Individual fragment may be too long to sequence completely

500 bp

Fig 2.8

A long sequence generated from smaller restriction fragments. Abbreviations: A, *Alu*I site; S, *Sau*3A site

Generating overlapping fragments with two different restriction endonucleases

In the first method restriction enzymes are used to obtain the overlapping fragments. Usually enzymes such as *Sau*3A or *Alu*I are used, as these recognize 4-nucleotide restriction sequences and so are likely to cleave the target DNA relatively frequently. A different set of restriction fragments are produced by each enzyme, the fragments obtained with the first enzyme overlapping those produced by the second (*Figure 2.8*). Each fragment must then be cloned back into an M13mp vector and single stranded DNA for sequencing prepared from the resulting clear plaques.

Although this is a popular means of obtaining overlapping fragments the approach suffers from the drawback that the restriction sites may be inconveniently placed and individual fragments may still be too long to be sequenced. In practice you may need to use four or more restriction enzymes in order to cover all the gaps. This is true even if you already know the positions of the restriction sites in your starting DNA and so do not waste time with enzymes that are completely unsuitable.

Producing random fragments

With restriction enzymes you are limited by the positions of the restriction sites in the DNA molecule. Clearly, a method that breaks the DNA at random positions would be better. Random breaks in a DNA molecule can be produced by either of two methods:

- by sonication
- by treatment with an endonuclease that cleaves at random positions

Sonication (Bankier *et al.* 1987) makes use of specialized equipment that generates high frequency vibrations from the tip of a metal probe. If the probe is placed in a DNA solution then the ultrasonic vibrations randomly shear the DNA molecules. The technique has the advantage that the lengths of the resulting DNA fragments are dependent on the duration and frequency of the treatment, so it is easy to obtain fragments of defined lengths (e.g. in the range

300–600 bp). A disadvantage is that the ends of the DNA fragments may be ragged (i.e. they may have short 5′ or 3′ overhangs) so further enzymatic treatment (for instance, with exonuclease VII) is needed to produce blunt-ended molecules that can be ligated into the cloning vector.

The second approach to production of random fragments makes use of deoxyribonuclease I (DNase I) which, under the correct conditions, makes random double stranded cuts in DNA molecules. Unfortunately, DNase I activity is difficult to control, especially as the potency of the enzyme declines during storage in the freezer. A calibration experiment therefore has to be carried out each time the procedure is used, to make sure the extent of DNA breakage is just right. This makes it less popular than sonication, the latter being highly reproducible.

◇ To obtain double stranded cuts, Mn⁺⁺ ions must be present in the DNase I reaction mixture. If the Mn⁺⁺ ions are replaced with Mg⁺⁺ then the enzyme makes single stranded nicks.

Progressive deletions

This is the best strategy for sequencing a long piece of DNA. The idea is that the starting DNA is progressively shortened by deleting nucleotides from one end. DNA molecules are removed from the reaction at different stages in the deletion process, so you end up with a family of molecules of different lengths. The sequence at the deleted end of each molecule is then determined. If you have carried out the procedure correctly then each DNA sequence runs into the next one, providing you with the required overlaps (*Figure 2.9*). Unlike the previous two methods, progressive deletions enable you to build up the master sequence gradually from one end of the molecule, without any gaps. This means that you can curtail the project as soon as you have obtained the piece of sequence (a gene, for example) that you are interested in, thus saving time.

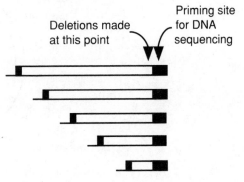

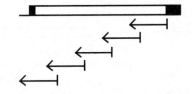

Fig 2.9

A long DNA sequence generated from a family of deleted molecules

5' overhangs are
susceptible to ExoIII

Bam HI   -N-N-G
             -N-N-C-C-T-A-G

Xba I   -N-N-T
             -N-N-A-G-A-T-C

3' overhangs are resistant

Pst I   -N-N-C-T-G-C-A
            -N-N-G

Fig 2.10

Overhangs resulting from restriction with
different enzymes. N, any nucleotide

◇ ExoIII activity is terminated by adding
EDTA, which chelates Mg^{++} ions that the
enzyme needs in order to work.

Several ways of generating the progressive deletions have been developed but the most favoured method makes use of exonuclease III (ExoIII; Guo and Wu 1983). This enzyme is particularly suitable for the task as it digests double stranded DNA only from a terminus with a 5' overhang. A 5' overhang is left after restriction with any one of several enzymes, for example *Bam*HI and *Xba*I. ExoIII does not digest DNA from a terminus with a 3' overhang, the product of restriction with, amongst others, *Pst*I (*Figure 2.10*). The procedure is therefore to restrict the M13 RF with a combination of *Pst*I and *Bam*HI or *Xba*I. Take another look at *Table 2.1* and ask yourself what this results in if you are working with a fragment of DNA inserted into the *Eco*RI site of any of the M13mp vectors from number 8 upwards. The answer is that you produce a linear RF molecule with a 5' overhang at one end and a 3' overhang at the other. The 5' overhang, the one that is susceptible to ExoIII digestion, is adjacent to the cloned DNA, so the linear molecule can be used as the substrate for your progressive deletions (*Figure 2.11*).

It is important to carry out control experiments to determine the rate at which the enzyme shortens the DNA, so you know the appropriate time to leave the reaction between removal of samples for sequence analysis (*Figure 2.12*). You must also terminate the ExoIII activity as rapidly as possible after removing a sample, otherwise the enzyme continues working and the deletions are longer than you

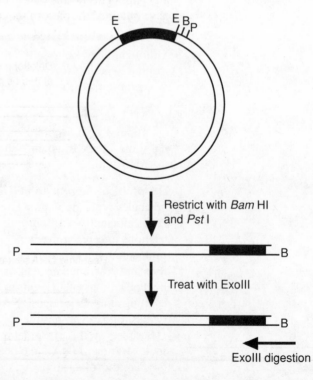

M13mp18 with DNA inserted
in to the *Eco*RI site

Restrict with *Bam* HI
and *Pst* I

Treat with ExoIII

ExoIII digestion

Fig 2.11

Producing a linear M13 molecule suitable
for ExoIII digestion. Abbreviations: B,
*Bam*HI site; E, *Eco*RI site; P, *Pst*I site

Fig 2.12

The results of a control experiment to determine the rate at which ExoIII shortens a DNA molecule. Approximately 1 μg of treated DNA has been loaded per track. The starting molecule is made up of 3 kb of vector sequence plus 3 kb of inserted DNA. Comparison with the markers (M) shows that treatment with ExoIII is shortening the molecule at a rate of approximately 400 bp/min. Samples taken every minute would therefore provide a family of molecules with lengths suitable for generating a long, overlapping sequence. (Reproduced with permission from Heinrich, P. (1991) in Ausubel, F.M. et al. (ed.) (1994). *Current protocols in molecular biology*. Greene Publishing Associates and John Wiley and Sons, New York, page 7.2.5 ©1991.)

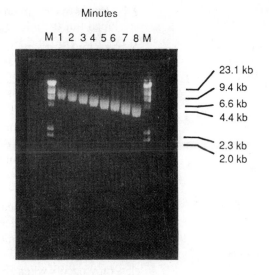

Minutes

M 1 2 3 4 5 6 7 8 M

23.1 kb
9.4 kb
6.6 kb
4.4 kb

2.3 kb
2.0 kb

expect. In fact the major problem with the technique is that it can be very difficult to control the reaction.

The samples removed at different time points are treated with exonuclease VII so that each molecule has two blunt ends. The molecules are then recircularized with DNA ligase, cloned, and single stranded DNA for sequencing prepared from the clear plaques that you obtain.

2. Other ways of obtaining single stranded DNA

Although the M13mp series of vectors are very versatile they are not absolutely perfect for production of single stranded DNA. There are two commonly acknowledged problems:

- M13 vectors are inefficient at cloning fragments greater than about 3 kb in length
- some DNA fragments shorter than 3 kb cannot be cloned in M13 at all

The problems appear to derive from the way in which M13 DNA molecules replicate within *E. coli* host cells. The replication system works efficiently with normal M13 genomes but tends to introduce deletions and rearrangements into genomes carrying inserted DNA fragments longer than a few kb. Shorter fragments also suffer deletions and rearrangements if they contain particular sequences which the replication system seems to be unable to handle. In either case the resulting single stranded DNA (if any is produced at all) cannot be used as it no longer has the correct sequence.

Because of these problems alternative vectors for synthesis of single stranded DNA have been developed. The most successful of these are the 'phagemids'.

2.1 Phagemids: phage–plasmid hybrids

◇ For a more detailed description of the pEMBL vectors, a popular group of phagemids, see Dente et al. (1983).

A phagemid is essentially a plasmid cloning vector and as such it replicates inside *E. coli* in exactly the same way as a plasmid such as pUC18. Plasmid replication systems do not suffer to such a great extent from the instabilities of the M13 system, so pieces of DNA longer than 3 kb can be cloned without too much difficulty.

If we have cloned our DNA in a plasmid then how do we obtain single stranded DNA? Phagemids in fact carry two replication origins, one being the standard plasmid origin that is used for replication of the double stranded DNA molecules, the other derived from M13 (or a related single stranded phage such as f1). The phage origin enables a single stranded version of the genome to be made. Note however that the phagemid does not itself carry any of the M13 genes, so on its own cannot make use of the M13 origin of replication. Synthesis of single stranded DNA, and release of phage particles, only occurs when the transformed cells are 'superinfected' with a 'helper phage'. The helper phage carries the genes for the M13 replicative enzymes and coat proteins, and is able to convert the phagemids into single stranded DNA molecules, which are assembled and secreted from the cell. The procedure is illustrated in *Figure 2.13*.

◇ The only major problem with phagemids is that sometimes the yield of single stranded DNA is low compared with a conventional M13 vector.

The list of phagemids is growing all the time as most of the new vectors that come on to the market incorporate an M13 origin of replication. Their convenience as all-round cloning systems means that they may soon supersede standard M13 vectors altogether.

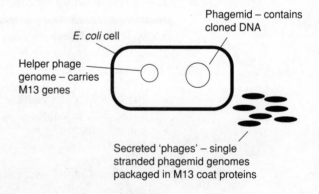

Fig 2.13

Secretion of M13 'phage' particles from a cell containing a phagemid and helper phage genome

2.2 Obtaining single stranded DNA by the polymerase chain reaction

Until the late 1980s cloning, in an M13 or phagemid vector, was the only way of obtaining single stranded DNA for chain termination sequencing. Nowadays, an alternative is provided by the polymerase chain reaction (PCR). A polymerase chain reaction results in a large

◇ For a full description of PCR, including its numerous applications, see Erlich (1989).

number of identical double stranded DNA fragments, generated by amplification of minute amounts of starting material. The starting material can be DNA from any source—chromosomal DNA, viral DNA, DNA inserted in a cloning vector, whatever you like—and as little as a single molecule is all that is needed.

The one limitation with PCR is that the reaction must be primed by two oligonucleotides that anneal to the target molecule and delineate the region to be amplified. The sequences of the two oligonucleotide primers must be complementary to the target molecule, otherwise they cannot anneal. To carry out a PCR you must therefore know at least some of the sequence of the target DNA. Note, though, that the sequence need only be known for the boundary regions. The internal segment of DNA, which forms the bulk of each amplified fragment, can be completely uncharacterized. PCR is therefore an excellent way of obtaining DNA from regions that remain as gaps in a master sequence being built up from overlapping fragments.

How do we sequence DNA produced by PCR? The DNA is double stranded so is not itself suitable for chain termination

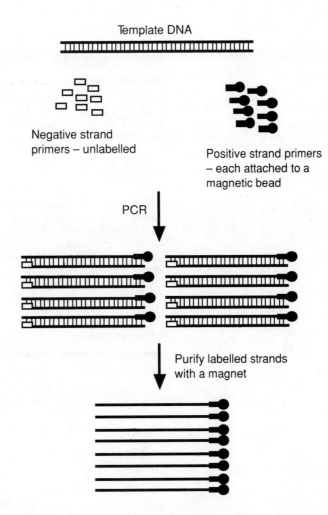

Fig 2.14

Using magnetic labelling as a means of obtaining single stranded DNA after PCR

sequencing (except by special methods—see Chapter 3, Section 2.2.3). One possibility would be to clone the DNA into an M13 or phagemid vector and obtain single stranded molecules in the normal way. However, this is time-consuming. A quicker way is to produce single stranded DNA more directly from the amplified DNA. There are two possibilities here:

1. You could carry out the initial PCR with one normal and one modified primer, the modified primer altered in such a way that the DNA strands synthesized from it are easily purified. A clever way of doing this is by attaching small magnetic beads to one of the primers. After the PCR, single stranded DNA can be obtained by separating the 'magnetic' strand from the ordinary strand (*Figure 2.14*). A similar method makes use of biotin-labelled primers, with single strands being separated by binding to avidin, a protein that has a high affinity for biotin.

2. Alternatively, you can carry out a second PCR with a large excess of one of the primers. This results in over-amplification of one strand. You do not obtain absolutely pure single stranded DNA, but the final mixture is sufficiently enriched in the over-amplified strand to be usable in a chain termination experiment.

Further reading

Brown, T.A. (ed.) (1991). *Molecular biology labfax*. BIOS Scientific Publishers, Oxford—a source of *E. coli* genotypes, descriptions of cloning vectors, etc.

Messing, J. and Bankier, A.T. (1989). The use of single-stranded DNA phage in DNA sequencing. In *Nucleic acids sequencing: A practical approach*, (ed. C.J.Howe and E.S.Ward), pp. 1–36. IRL Press at Oxford University Press—this is currently the best detailed description of the use of M13 and phagemid vectors in DNA sequencing.

Pouwels, P.H. (1991). Survey of cloning vectors for *Escherichia coli*. In *Essential molecular biology: A practical approach*, Vol. I, (ed. T.A.Brown), pp. 179–239. IRL Press at Oxford University Press—includes a description of different kinds of M13 and phagemid vectors.

Towner, P. (1991). Purification of DNA. In *Essential molecular biology: A practical approach*, Vol. I, (ed. T.A.Brown), pp. 47–68. IRL Press at Oxford University Press—provides details on methods for purifying DNA from M13 and phagemid vectors.

References

Bainbridge, B.W. (1991). Microbiological techniques for molecular biology: bacteria and phages. In *Essential molecular biology: A practical approach*, Vol. I, (ed. T.A.Brown), pp. 13–45. IRL Press at Oxford University Press.

Bankier, A.T., Weston, K.M., and Barrell, B.G. (1987). Random cloning and sequencing by the M13/dideoxynucleotide chain termination method. *Methods in Enzymology*, **155**, 51.

Birnboim, H.C. and Doly, J. (1979). A rapid alkaline method for screening recombinant plasmid DNA. *Nucleic Acids Research*, **7**, 1513.

Dente, L., Cesareni, G., and Cortese, R. (1983). pEMBL: a new family of single stranded plasmids. *Nucleic Acids Research*, **11**, 1645.

Erlich, H.A. (ed.) (1989). *PCR technology: Principles and applications for DNA amplification*. Stockton Press, New York.

Guo, L.-H. and Wu, R. (1983). Exonuclease III: use for DNA sequence analysis and in specific deletions of nucleotides. *Methods in Enzymology*, **100**, 60.

Horwitz, J.P. *et al.* (1964). Substrates for cytochemical demonstration of enzyme activity. I. Some substituted 3-indolyl-β-D-glycopyranosides. *Journal of Medical Chemistry*, **7**, 574.

Marvin, D.A. and Wachtel, E.J. (1975). Structure and assembly of filamentous bacterial viruses. *Nature*, **253**, 19.

Messing, J., Crea, R., and Seeburg, P.H. (1981). A system for shotgun sequencing. *Nucleic Acids Research*, **9**, 309.

Ullman, A. and Perrin, D. (1970). Complementation in β-galactosidase. In *The lactose operon*, (ed. J.R.Beckwith and D.Zipser), pp. 143–72. Cold Spring Harbor Laboratory, Cold Spring Harbor, New York.

Yanisch-Perron, C., Vieira, J., and Messing, J. (1985). Improved M13 phage cloning vectors and host strains. Nucleotide sequences of the M13mp18 and pUC19 vectors. *Gene*, **33**, 103.

3 Chain termination sequencing: carrying out the strand synthesis reactions

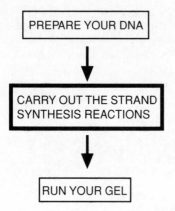

PREPARE YOUR DNA

↓

CARRY OUT THE STRAND
SYNTHESIS REACTIONS

↓

RUN YOUR GEL

You have prepared your single stranded DNA template and are now ready to carry out the strand synthesis reactions. These will provide you with the four families of chain terminated polynucleotides that you will subsequently load on to the polyacrylamide electrophoresis gel.

The strand synthesis reactions are the easiest part of the DNA sequencing procedure. Although a number of different reagents have to be mixed together the manipulations are not much more complicated than setting up a restriction digest. As in all molecular biology procedures, if you use good quality reagents and carry out the pipettings accurately then the experiment will almost always work. The DNA sequencing reactions are made even easier by the availability of commercial sequencing kits, which contain premixed reagents and provide precise instructions on how they should be used. In fact, reading skills are not even required as in most kits the reagents are colour-coded, so it becomes merely a matter of adding 2 µl from the blue tube to 5 µl from the red tube. This sounds great, but if you are starting out in sequencing you should be very wary of placing too much reliance on kits. Mixing together reagents from different coloured tubes has to be supplemented with thought and understanding if you are to use DNA sequencing to its full advantage. If you do not understand what the blue tube does, or what is happening when it says 'place in a 37°C water bath for 5 minutes', then you will have little idea of what to do when things go wrong. If you do not know what to do when things go wrong then your research will quickly reach a full stop.

We will begin this chapter by looking in turn at each of the reagents used in the strand synthesis reactions.

1. The components of the strand synthesis reactions

First we must remind ourselves of the events that take place during the strand synthesis reactions (*Figure 3.1*). As described in Chapter 1, Section 2.1.1, the first step is to anneal a short oligonucleotide primer (described in Section 1.1 of this chapter) to each single stranded template molecule. The primer acts as the starting point for synthesis of a new polynucleotide chain, complementary to the template strand. The strand synthesis is carried out by a DNA polymerase enzyme (Section 1.2) and requires deoxynucleotide triphosphates (Section 1.3) as substrates. One or more of these dNTPs is labelled (Section 1.4), usually with a radioactive marker, so that the results of the gel electrophoresis can later be visualized by autoradiography or some other means. Strand synthesis is not allowed to continue to completion as a chain terminating nucleotide (Section 1.5) is included in the reaction mixture. Four families of chain terminated molecules are produced, one family consisting of polynucleotides that terminate in A, a second family of C-terminated polynucleotides, and so on. The four families are then fractionated by polyacrylamide gel electrophoresis to obtain the banding pattern from which the sequence is read.

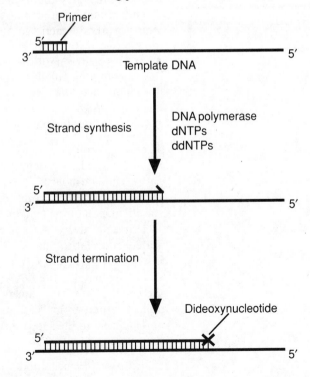

Fig 3.1

The strand synthesis and termination events that underlie chain termination sequencing

1.1 The primer

◇ See the beginning of Section 1.2 for the definition of 'template-dependent'.

The primer is needed because template-dependent DNA polymerases are unable to initiate DNA synthesis on an entirely single stranded molecule. A short double stranded region is needed to provide a 3′

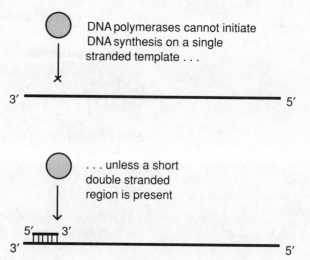

Fig 3.2

Why the primer is needed

end on to which the polymerase can add new nucleotides (*Figure 3.2*). The primer provides this 3' end: without it strand synthesis cannot take place.

You might at first imagine that the primer is a bit of a nuisance in a DNA sequencing experiment, an extra reagent to obtain and an extra step to worry about. In fact the primer plays a critical role in the chain termination procedure as it delineates the starting point for the strand synthesis reactions. As a result the primer ensures that:

- all the strands synthesized by the DNA polymerase have one end in common, which is essential if the lengths of the terminated chains are to reflect the sequence of the template strand (see *Figure 1.4B*)
- only the desired part of the template molecule is sequenced

The second point is important because you cannot obtain more than a few hundred base pairs of sequence in a single experiment. The entire template molecule—vector plus inserted DNA—is several kilobases in length, so you must direct each sequencing experiment at the specific region of the template that you are interested in. In practice, you virtually always use a primer that anneals to the template immediately adjacent to one of the junctions between the vector and the inserted DNA (*Figure 3.3*). This is the ideal position at which to begin strand synthesis as the DNA polymerase immediately starts to read the insert DNA. The sequence that you obtain is therefore derived almost entirely from the cloned DNA fragment.

The priming site shown in *Figure 3.3* is within the *lacZ'* region of the vector molecule. This region has the same sequence in all vectors

Fig 3.3

The primer anneals to the template at a position immediately adjacent to the inserted DNA

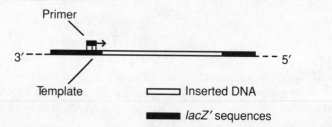

that carry the *lacZ'* gene (i.e. all M13 and phagemid vectors) and so it is not necessary to synthesize a different primer for each sequencing experiment. A single 'universal' primer, whose sequence is complementary to the relevant segment of the *lacZ'* gene, is suitable for all experiments regardless of the origins of the cloned DNA.

Strand synthesis can be initiated by quite short primers, theoretically as small as 2 nucleotides. However, it is important to ensure that the primer is specific for its target site and does not, by chance, also anneal to a second position in the template. If this happened then two sequences would be superimposed on the polyacrylamide gel, neither being readable. Most commercially available universal primers are 15–24 nucleotides (*Figure 3.4*), long enough to reduce the chances of double priming to an absolute minimum.

◇ An oligonucleotide 15 nucleotides in length is said to be a '15-mer'; one 20 nucleotides long is a '20-mer', and so on.

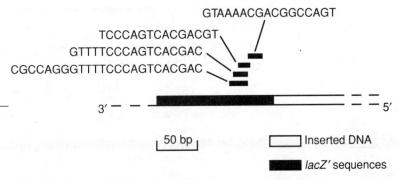

Fig 3.4

Sequences and annealing positions of four universal sequencing primers marketed by New England Biolabs. The same or similar sequencing primers are also sold by several other companies

Are there any occasions when you might use a primer other than a universal one? One possibility is to use a primer that anneals within the inserted DNA rather than to one side of it. This would enable you to obtain a sequence from a region of the inserted DNA that is too far away from the insert-vector junction to be sequenced with a universal primer. You could therefore use internal primers to obtain the entire sequence of a DNA insert that is too long to be sequenced in one go (*Figure 3.5*). Although this may sound like an attractive idea, the subcloning strategies described in Chapter 2, Section 1.4.3 are better options for obtaining long stretches of sequence. There are several disadvantages with internal primers, the main one for many labs being the cost of preparing a special oligonucleotide for each sequencing experiment.

◇ Each of the subclones generated by one of the strategies in Chapter 2, Section 1.4.3 would be sequenced with the universal primer.

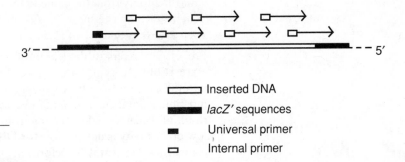

Fig 3.5

Using internal primers to build up a long sequence

1.2　The DNA polymerase

Any enzyme that synthesizes DNA is called a DNA polymerase. Most DNA polymerases are template-dependent, meaning that the new DNA strand is synthesized on an existing single stranded template molecule, the sequence of the new strand being complementary to the sequence of the template (*Figure 3.6*). Some enzymes use a DNA template, and so are DNA-dependent DNA polymerases, others use RNA. The RNA-dependent DNA polymerases are also called reverse transcriptases. All DNA polymerases synthesize DNA in the 5′ to 3′ direction, meaning that they add nucleotides on to the 3′ end of the strand that is being made (*Figure 3.6*). If synthesis is in the 5′→3′ direction then the template must be read in the 3′→5′ direction, as the two strands in a double stranded nucleic acid molecule are always antiparallel.

New strand being synthesized
by DNA polymerase

$$5'--A-T-C-T-T-A-T-G-C \longrightarrow$$
$$\quad\ \ |\ \ |\ \ |\ \ |\ \ |\ \ |\ \ |\ \ |\ \ |$$
$$3'--T-A-G-A-A-T-A-C-G-C-C-T-A-A--5'$$

Fig 3.6

Template-dependent DNA synthesis

Template strand

　All living organisms possess DNA polymerases, these enzymes playing essential roles in the replication and repair of cellular DNA molecules. In addition, many bacteriophages and viruses have genes that, after infection, cause the cell to synthesize new DNA polymerases that replicate the phage or viral DNA. There are therefore many different DNA polymerases, with subtly different properties, to choose from.

　To be suitable for use in DNA sequencing a DNA polymerase must satisfy a number of criteria. The most important of these are high processivity and low exonuclease activity. In the next few pages we will discover what is meant by these two terms, and also look at other, less critical, properties that the 'ideal' DNA sequencing enzyme should possess. We will then examine the range of DNA polymerases that are used in DNA sequencing and assess the good and bad points of each one.

1.2.1　Processivity

Processivity refers to the length of new strand that a DNA polymerase is able to synthesize before the reaction terminates through natural causes. No DNA polymerase is able to synthesize DNA indefinitely, and chain termination always occurs at some point as a result of the enzyme dissociating from the template. An enzyme with low processivity is not suitable for DNA sequencing as it may dissociate from the template before incorporating a chain terminating nucleotide. The randomly terminated strands produced by dissociation result in extra bands in the polyacrylamide gel, making it difficult if not impossible to read the sequence. Sequencing enzymes must therefore have high processivities.

1.2.2 Exonuclease activities

Many DNA polymerases are dual function enzymes, being able to degrade DNA as well as synthesize it. This may seem peculiar but fits in with the roles these enzymes play in the cell. DNA polymerase I of *E. coli*, for example, is a DNA repair enzyme, which means that it has to be able to remove nucleotides from a DNA strand in order to correct mistakes made during DNA replication or arising from mutation. To do this the enzyme needs a 5'→3' exonuclease activity (*Figure 3.7A*). In addition, DNA polymerase I, like most DNA

A. The 5'→3' exonuclease enables a DNA
polymerase to correct an error in an existing strand

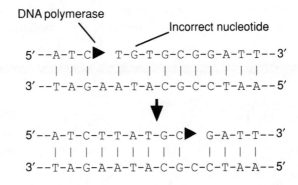

B. The 3'→5' exonuclease enables DNA
polymerase to correct its own mistakes

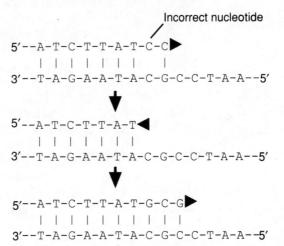

Fig 3.7

The exonuclease activities of DNA
polymerase I

polymerases, has a proof-reading function that helps it correct any mistakes that it makes during DNA synthesis. This requires a 3′→5′ exonuclease activity, as the enzyme has to go into reverse in order to remove incorrect nucleotides from the DNA strand it is synthesizing (*Figure 3.7B*). The two exonuclease activities are therefore essential for the correct functioning of the 'polymerase' enzyme.

Unfortunately, both exonuclease activities are a hindrance during chain termination sequencing. The 5′→3′ exonuclease, in particular, is a problem as enzymes that possess this activity can remove nucleotides from the 5′ ends of the chain terminated strands (*Figure 3.8*). If this happens then the lengths of the chain terminated molecules become variable and no longer reflect the sequence of the template. The banding pattern on the polyacrylamide gel will therefore not be interpretable. To be suitable for chain termination sequencing a DNA polymerase must therefore have a zero or negligible 5′→3′ exonuclease activity.

The 3′→5′ exonuclease is also a problem, though not to such a great extent. The proofreading function provided by this exonuclease activity is a nuisance as you may wish to include modified nucleotides such as deoxyinosine triphosphate (dITP) in your sequencing reactions. These modified nucleotides are used under special circum-

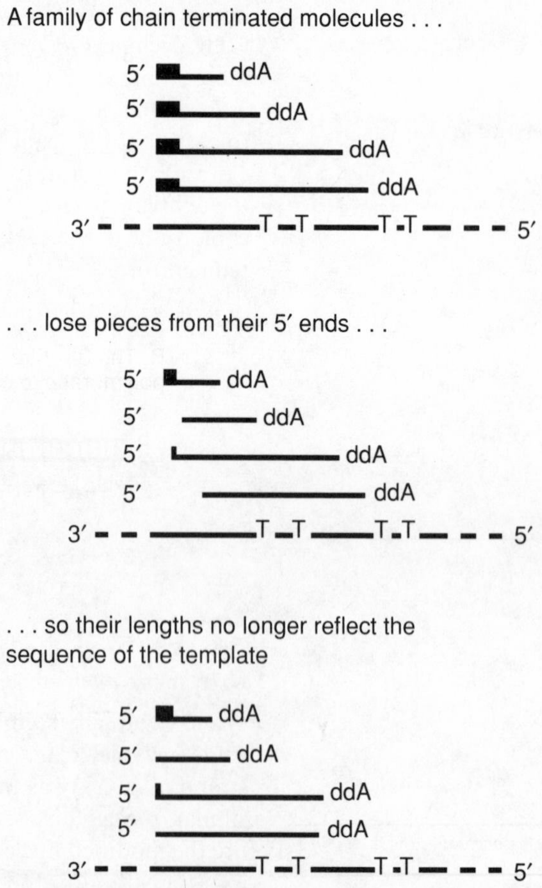

A family of chain terminated molecules . . .

. . . lose pieces from their 5′ ends . . .

. . . so their lengths no longer reflect the sequence of the template

Fig 3.8

The problem caused by a 5′→3′ exonuclease activity

stances to improve the results of the sequencing experiment. The problem is that the proof-reading functions of many polymerases are able to recognize the modified nucleotides and remove them whenever they are incorporated into the growing polynucleotide chain. This defeats the purpose of using the modified nucleotides in the first place. The proof-reading function may even be able to prevent chain termination by removing the chain terminating nucleotide as soon as it is added. On top of all this there are other problems that are generally ascribed to the $3' \rightarrow 5'$ exonuclease. Our ideal polymerase for DNA sequencing should therefore be exonuclease free.

1.2.3 Other features of the ideal polymerase for DNA sequencing

As well as high processivity and low exonuclease activity, the ideal DNA polymerase for chain termination sequencing should also have the following features.

Activity at 55°C and above

A frequent problem in a sequencing experiment is that parts of the template are inaccessible because of base pairing that occurs between different regions of the single stranded molecule. The base pairing results in secondary structures, such as hairpin loops, which the DNA polymerase cannot penetrate (*Figure 3.9*). Strand synthesis therefore becomes stalled. This can be especially troublesome if the insert DNA is GC-rich, as G-C base pairs are stronger than A-T ones and so are more stable at 37°C, the optimal temperature for most DNA polymerases. One possible solution to this secondary structure problem is to carry out the sequencing reactions at 55°C, or even higher, as the elevated temperature destabilizes the base pairing and opens up the template to the DNA polymerase. Clearly, this solution is practicable only if the DNA polymerase is able to work at the higher temperature.

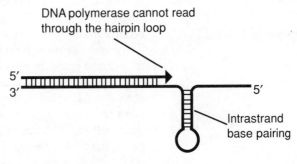

DNA polymerase cannot read through the hairpin loop

5'
3'

5'

Intrastrand base pairing

Fig 3.9

The problem caused by a hairpin loop in the template DNA

Ability to use modified nucleotides as substrates

We have already seen how the proof-reading exonuclease activity of some enzymes can remove modified nucleotides from the growing strand as quickly as the polymerase adds them in (Section 1.2.2). A similar, though distinct, problem arises if the polymerase is unable to use the modified nucleotides as substrates, and so does not add them into the strand in the first place. Some DNA polymerases are very

◇ The use of modified nucleotides such as dITP in chain termination sequencing will be covered in Section 1.3.

scrupulous about the the types of nucleotide that they use, and may be unable to recognize ddNTPs, dITP and other modified versions. These enzymes are no good for DNA sequencing.

Rate of strand synthesis

The rate at which a DNA polymerase synthesizes new DNA is important because we do not want to have to incubate the strand synthesis reactions for an inordinately long time. Most DNA polymerases have average synthesis rates of tens of nucleotides per second, and so are theoretically able to polymerize 500 nucleotides in less than one minute. Even this may, in practice, be too slow if the template contains several regions which, because of the nucleotide sequence, cause the polymerase to slow down. Our ideal sequencing polymerase should therefore have an elongation rate of hundreds of nucleotides per second.

1.2.4 DNA polymerases used in DNA sequencing

We have established that the 'ideal' DNA polymerase for chain termination sequencing must have high processivity and low exonuclease activity, must remain active at elevated temperatures, must be able to use modified nucleotides as substrates, and must have an elongation rate of at least several hundred nucleotides per second. Does such an enzyme exist?

In fact there are several distinct DNA polymerases that satisfy enough of these criteria to be used regularly in DNA sequencing. One of these enzymes, the modified form of T7 DNA polymerase called Sequenase, comes very close to being the 'ideal' enzyme. We will look first at Sequenase and then consider the merits of the other enzymes.

Sequenase

◇ Sequenase is marketed by Amersham International.

Sequenase is a modified version of the DNA polymerase encoded by bacteriophage T7. The enzyme is responsible for replicating the T7 genome and is synthesized by *E. coli* cells shortly after infection with the phage. The unmodified T7 DNA polymerase is highly processive, on average polymerizing 2000–3000 nucleotides before dissociating from its template, has an elongation rate of about 300 nucleotides per second and is able to use many modified nucleotides as substrates. In addition, it retains some activity at 55°C and has no 5′→3′ exonuclease activity. The perfect enzyme? Not quite, as the unmodified polymerase has a potent 3′→5′ exonuclease activity that makes it a poor choice for chain termination sequencing.

It is possible to suppress the 3′→5′ exonuclease of the unmodified enzyme by using special reaction conditions, but this does not completely solve the problem. Instead, the answer is to modify the enzyme so that the exonuclease activity is inactivated. There are two ways of doing this:

● by chemical modification—oxidation—of the enzyme, which when carried out in the correct way reduces the exonuclease activity by 99 per cent

- by genetic engineering, which deletes a segment of the enzyme's polypeptide chain and almost totally eliminates the exonuclease activity

The chemical modification results in Sequenase Version 1 (Tabor and Richardson 1987*a*), and genetic engineering in Version 2 (Tabor and Richardson 1989*a*). Both are excellent enzymes for chain termination sequencing, allowing 500 bp of sequence to be obtained in a single experiment and enabling you to use all the available tricks with modified nucleotides. We will encounter Sequenase again when we run through the sequencing procedures.

◇ The sequencing procedures are described in Section 2 of this chapter.

Klenow polymerase

Sequenase enzymes have been available only since the late-1980s. Before then virtually everyone used the Klenow polymerase for chain termination sequencing (Sanger *et al.* 1977). The Klenow polymerase is another modified enzyme, this time derived from DNA polymerase I of *E. coli*. Unmodified DNA polymerase I is not suitable for DNA sequencing because of its $5' \rightarrow 3'$ exonuclease, but this activity can be separated from the polymerase by cleaving the enzyme into two segments. The correct name for the Klenow polymerase is, in fact, 'Klenow fragment of DNA polymerase I', which indicates where it comes from. Klenow polymerase was originally prepared by proteolytic cleavage of the parent enzyme, but as with Sequenase is now usually obtained by genetic engineering.

◇ The Klenow polymerase also has a $3' \rightarrow 5'$ exonuclease activity, but this is relatively weak and is generally not a problem in sequencing experiments.

Klenow polymerase served us well for many years and is still frequently used today. Unfortunately it has its disadvantages, the major one being that a low processivity and low elongation rate combine to make it difficult to obtain more than about 250 bp of sequence in one experiment. In addition, sequences obtained with Klenow polymerase tend to have higher backgrounds (i.e. faint bands in all tracks) than seen with Sequenase, and ambiguities are more frequent. Fortunately, these ambiguities are not random and the correct nucleotide sequence at any position can usually be determined. We will return to this point when we deal with sequence reading in Chapter 4, Section 2.3.2.

Thermostable DNA polymerases

If you have carried out PCR experiments then you will be familiar with thermostable enzymes such as *Taq* DNA polymerase. These enzymes are obtained from bacteria that live at high temperatures in environments such as hot springs and thermal vents. *Taq* DNA polymerase has an optimal temperature of 72°C and other thermostable enzymes (e.g. Vent DNA polymerase) are active at even higher temperatures.

Thermostable enzymes are often used in DNA sequencing as their ability to work at high temperatures makes them ideal for sequencing templates that can form strong base paired structures (see *Figure 3.9*). Neither Sequenase nor Klenow can work at anything approaching the temperatures tolerated by these thermostable polymerases.

The enzymes are also used in a special type of chain termination sequencing, called the thermal cycle procedure, which is becoming popular for sequencing DNA fragments that have been obtained by PCR amplification (Section 2.2.3).

Taq DNA polymerase is currently the first choice thermostable enzyme for sequencing (Innis *et al.* 1988). The native form has a strong 5′→3′ exonuclease activity but, as we have now come to expect, this can be deleted by genetic engineering. Vent DNA polymerase is significantly more thermostable than *Taq* but has a lower processivity; again there are modified versions that lack exonuclease activities.

1.3 The deoxynucleotide triphosphates

As we all know, DNA is synthesized by polymerization of deoxynucleotide triphosphates (dNTPs). In a chain termination sequencing experiment you therefore include each of the four standard dNTPs (dATP, dCTP, dGTP, dTTP) in the reaction mixtures. But what about the 'modified nucleotides' that have drifted in and out of the discussion over the last few pages, what are they used for?

Two modified nucleotides are commonly used in chain termination sequencing (Mills and Kramer 1979; Barr *et al.* 1986):

- deoxyinosine triphosphate (dITP; *Figure 3.10A*)
- 7-deaza-dGTP (*Figure 3.10B*)

Both are modified versions of dGTP and both are used when the sequence of the template is such that intrastrand hairpin loops might form. We have already seen the consequences of hairpin loops in the template itself (see *Figure 3.9*) and noted that we may wish to carry out the strand synthesis reactions at 55°C or even 72°C in order to disrupt the base pairing. This is fine as far as the template is concerned, but what about the chain terminated strands that are subsequently synthesized? These new strands are complementary to the template which means that they too have the ability to form hairpin loops. This becomes a major problem when the chain terminated strands are run in the polyacrylamide gel as hairpin loops within a DNA molecule can affect its electrophoretic mobility. Instead of the bands being nicely spaced out some run too fast or too slow, leading to 'compressions' within which the DNA sequence cannot be read.

The modified nucleotides reduce the ability of the chain terminated molecules to form hairpin loops. Neither dITP nor 7-deaza-dGTP is able to form stable base pairs with C residues. If one or other of these modified nucleotides is used in place of dGTP in the strand synthesis reactions then the resulting molecules will not be able to form intrastrand 'G'-C base pairs. This means that only A-T base pairs can form, and under normal circumstances these on their own are not strong enough to result in stable hairpin loops. Compressions in the sequencing gel can therefore often be avoided by using dITP or 7-deaza-dGTP.

A. Deoxyinosine triphosphate

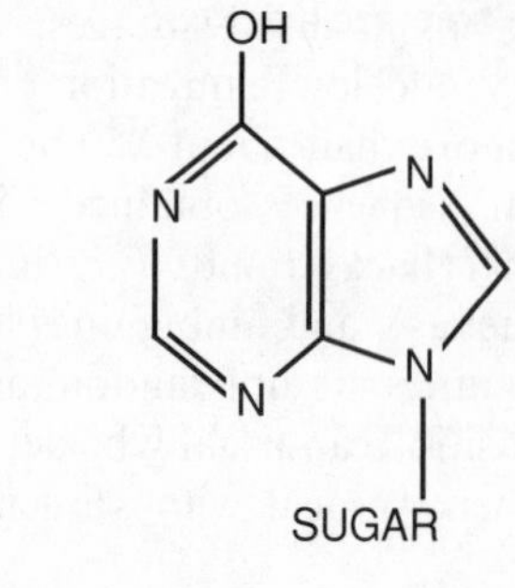

B. 7-deaza-dGTP

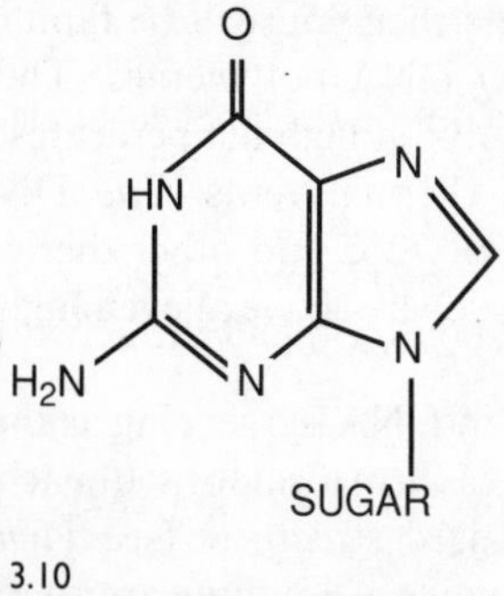

Fig 3.10

dITP and 7-deaza-dGTP

◇ We will find out more about compressions when we learn how to read sequencing gels in Chapter 4, Section 2.3.1.

Sequenase, Klenow, and Vent DNA polymerase are all able to use dITP and 7-deaza-dGTP as substrates, though with varying efficiencies; *Taq* DNA polymerase can use 7-deaza-dGTP but not dITP. In fact, dITP is usually the first choice if compressions are a problem, as this nucleotide tends to give better results than 7-deaza-dGTP. However, the modified nucleotides are not used routinely as both reduce the overall sharpness of the banding pattern in the sequencing gel.

1.4   The labelled nucleotide

After you have run your chain terminated strands in the sequencing gel you obviously need a way of seeing where the bands are. With an agarose gel you usually stain with ethidium bromide and visualise the bands on a u.v. transilluminator. The ethidium bromide binds to the DNA in the gel and then fluoresces in response to the u.v. radiation, revealing the positions of the bands.

◇ Refer to Chapter 4, Section 1.3 for a full description of the autoradiography step.

If you stain your sequencing gel with ethidium bromide then you might see a few bands, but the sensitivity and resolution will not be sufficient for you to make out much of the sequence. Instead, we usually include a radiolabelled nucleotide in the strand synthesis reactions and visualize the banding pattern by autoradiography (*Figure 3.11*).

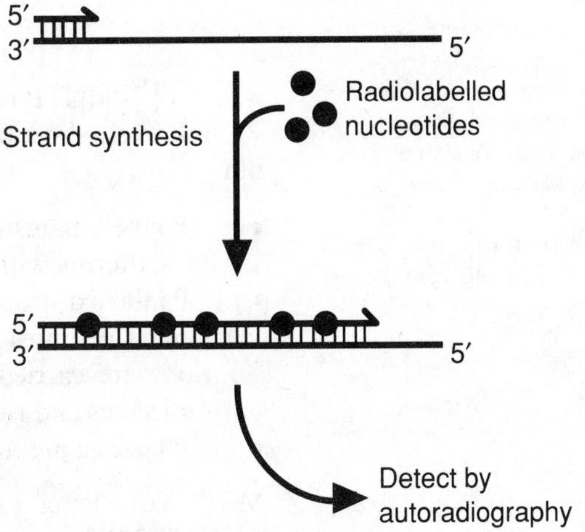

Fig 3.11

The traditional labelling-detection strategy used in chain termination sequencing

You will almost certainly use one of two types of labelled nucleotide in your sequencing reactions:

- a dNTP labelled with ^{32}P at the α position (*Figure 3.12A*)
- a dNTP labelled with ^{35}S by replacement of the O$^-$ attached to the α phosphorus (*Figure 3.12B*)

The main difference between the two types of radiolabelled nucleotide lies with the emission energies of the radioactive atoms. The β particles emitted by ^{32}P are approximately 10 times more energetic than those from ^{35}S. In practice, this means:

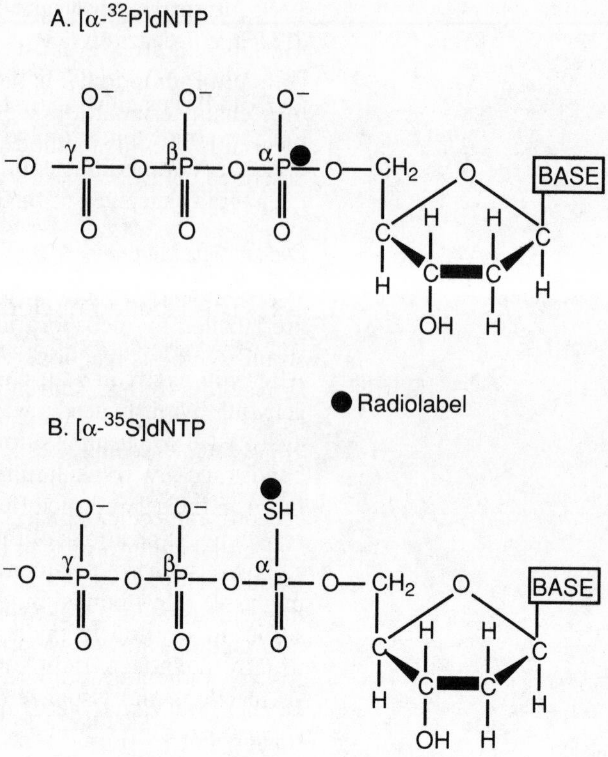

A. [α-^{32}P]dNTP

● Radiolabel

B. [α-^{35}S]dNTP

Fig 3.12

Labelled nucleotides

◇ In addition to ^{32}P and ^{35}S, ^{33}P-labelled nucleotides have recently become available. The emission energy of ^{33}P lies between the values for ^{32}P and ^{35}S (^{35}S, 0.167 MeV; ^{33}P, 0.249 MeV; ^{32}P 1.71 MeV). This means that ^{33}P displays the advantages of ^{35}S but requires a shorter exposure time.

- if a [^{32}P]dNTP is used the autoradiograph can be exposed for a relatively short time

but

- ^{32}P labelling results in fuzzier bands on the autoradiograph, due to scattering within the X-ray film
- ^{32}P labelled strands are unstable, which means that the polyacrylamide gel must be run immediately after the sequencing reactions are carried out—you cannot store the chain terminated molecules and run them another day
- ^{32}P poses a greater safety risk

Most sequencing experiments are therefore carried out with a [^{35}S]dNTP (Biggin *et al.* 1983). The autoradiograph must be exposed for longer but on the whole the results are better.

A labelled version of any of the four dNTPs could be used in the sequencing reactions. However, labelled dGTP would be a poor choice as at some stage you might wish to replace the dGTP with dITP or 7-deaza-dGTP. The best criterion to use when choosing a radiolabelled nucleotide is the GC content of the DNA you are sequencing. If the GC content is high then use labelled dCTP, if it is low then use labelled dATP. That way you maximize the amount of label that is incorporated into each chain terminated molecule, reducing the time needed for autoradiography.

1.4.1 Alternative labelling and detection strategies

The labelling strategy illustrated in *Figure 3.11*, followed by detection with autoradiography, is the standard way of visualizing the results of a chain termination sequencing experiment. Although you will invariably use this method there are alternatives, three of which are described below.

End-labelling the primer

In the standard labelling procedure the chain terminated molecules are labelled by incorporation of radioactive nucleotides during the strand synthesis reactions. An alternative is to attach a single radioactive atom to the 5′ end of the primer, before the primer is annealed to the template DNA. This is carried out with T4 polynucleotide kinase, which catalyses the transfer of a ^{32}P or ^{35}S atom from the γ position of an ATP donor to the 5′ end of a dephosphorylated acceptor molecule (*Figure 3.13*).

◇ Note that the donor molecule in the end- labelling reaction is ATP, not dATP. 'Dephosphorylated' means with the 5′ phosphate group removed.

End-labelling results in a chain terminated molecule that carries just a single radioactive atom, whereas in the standard method each molecule is multi-labelled. The molecules resulting from end-labelling therefore have lower radioactive activities, making it necessary to expose the autoradiograph for longer. So why use this strategy? In fact, we would generally use it only with special procedures for sequencing double stranded DNA, especially DNA obtained by PCR amplification. We will return to these procedures later in the chapter.

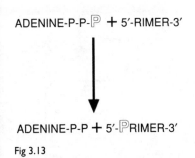

Fig 3.13

End-labelling a primer

Labelling with chemiluminescent tags

The hazards of radioactive labelling are well known and it is no surprise that non-radioactive alternatives are being sought. The drawback with many of the available non-radioactive methods is that they are not as sensitive as traditional radioactive labelling, and so cannot detect such small amounts of DNA. Great sensitivity is needed in DNA sequencing and to date non-radioactive labelling methods have not been widely used.

The most sensitive non-radioactive system so far developed uses chemiluminescence—the chemical production of light (Martin *et al.* 1991). The light emissions result from the enzymatic breakdown of a complex organic substrate called 'aryl-phosphate substituted 1,2-dioxetane'. The enzyme involved in this reaction—alkaline phosphatase—is too big to be attached to the DNA strand before electrophoresis so a two-stage process is used. The strand synthesis reactions are carried out with primers end-labelled with biotin and the chain terminated products run in the polyacrylamide gel as usual. The DNA bands are then blotted on to a nylon membrane, treated with a combination of reagents that bind alkaline phosphatase molecules to the biotin groups, and the dioxetane solution applied (*Figure 3.14*). The resulting flashes of light are detected on photographic film, producing an image similar to the autoradiograph obtained after radioactive labelling.

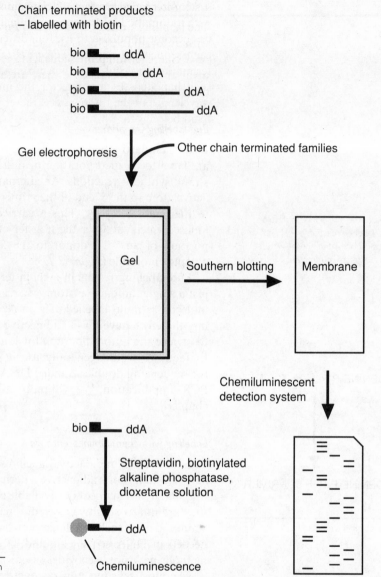

Chain terminated products
– labelled with biotin

bio ▬■▬ ddA
bio ■▬▬ ddA
bio ■▬▬▬ ddA
bio ■▬▬▬▬ ddA

Gel electrophoresis   Other chain terminated families

Gel   Southern blotting   Membrane

Chemiluminescent
detection system

bio ■▬ ddA

Streptavidin, biotinylated
alkaline phosphatase,
dioxetane solution

ddA

Chemiluminescence

Fig 3.14

Using a chemiluminescent labelling system
in chain termination sequencing

This technique is still new and it is not possible to say how widespread its applications will be in DNA sequencing. At the moment, radioactive labelling methods are used by virtually everyone, but this could easily change. The health and disposal hazards of radiochemicals are such that the non-radioactive alternatives may one day become mandatory.

Fluorescent labels

The final labelling strategy uses a fluorescent rather than chemiluminescent label (Prober *et al.* 1987). The fluorescence cannot be detected efficiently with photographic or X-ray films so a special imaging system is needed. In fact, fluorescent labels are currently used only with automatic DNA sequencing machines. One embellishment that has become popular is to use four different fluorescent labels, one for each sequencing reaction (*Figure 3.15*). The four families of chain terminated molecules are then mixed and fractionated in one lane. As each band passes the detector the imaging system determines the wavelength of the fluorescence, thereby identifying which family the band belongs to.

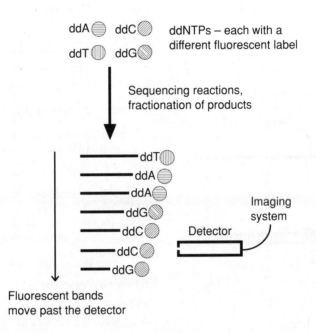

Fig 3.15

Using a fluorescent labelling system in chain termination sequencing

1.5   The chain terminating nucleotide

The introduction of dideoxynucleotides (ddNTPs) was one of the key innovations that made chain termination sequencing an 'efficient' procedure (Sanger *et al.* 1977), in contrast to the more haphazard methods it was derived from. A ddNTP lacks the hydroxyl group attached to the 3′ carbon of the sugar component of a normal deoxynucleotide (see *Figure 1.3*). The 3′ carbon participates in phosphodiester bond formation but without the hydroxyl group the reaction cannot take place. Incorporation of the ddNTP therefore blocks further strand elongation. The blockage is very effective and can be circumvented only by removal of the dideoxynucleotide from the DNA strand. If the sequencing enzyme lacks a 3′→5′ exonuclease activity then this removal is very unlikely to occur, so the chain stays terminated.

2. The sequencing reactions

You have prepared your template DNA and assembled the components of the sequencing reactions. Now you must mix them all together in that special way to make the chain terminated DNA strands.

There are several different procedures for carrying out the strand synthesis reactions. We will concentrate initially on the most commonly used method, the 'labelling-termination' procedure (Tabor and Richardson 1987*b*). Once we have sorted that out we can tackle some of the more specialized approaches.

2.1 The labelling-termination procedure

There are two main stages to this procedure (*Figure 3.16*):

1. Annealing the primer to the template.

2. Extending the annealed primer by strand synthesis to produce the chain terminated molecules.

The experiment can be set up in polypropylene microfuge tubes (0.5 or 1.5 ml sizes) or, if you intend to sequence several templates at once, in a 96-well microtitre tray. If you use a microtitre tray you should make sure it is the round bottomed variety, and you will need a special rotor on your bench-top centrifuge so you can spin the tray to mix the reagents.

2.1.1 Primer annealing

The first stage of any chain termination sequencing procedure is to anneal the oligonucleotide primer on to the template DNA molecule. When you carry out this step you aim to achieve two things:

- specificity
- efficiency

Specificity of priming

By 'specificity' we mean that the primer must anneal only to the desired position on the template, immediately upstream of the region that you wish to sequence. If primers also anneal at other positions then you will generate more than one set of chain terminated molecules, and will obtain two sequences at once (*Figure 3.17*). The two sequences will be superimposed in the polyacrylamide gel and the whole thing will be an unreadable mess.

Why might primers anneal to secondary positions on the template? Annealing depends on formation of base pairs between the primer and the template, so any template sequence that is complementary to the primer could act as an annealing site. Most primers are at least 15 nucleotides in length, making it unlikely that the template contains a second nucleotide sequence exactly the same as the desired priming site. But there is a good chance that other

1. Primer annealing

2. Strand synthesis

Fig 3.16

The two steps in the labelling-termination procedure

◇ A sequence generated from a secondary priming site is usually weaker than the main sequence, so gives 'ghost' bands on the autoradiograph. Sometimes these ghosts are weak enough for the real sequence to be read without too much trouble. But this is still a problem that is best avoided.

Fig 3.17

The problem caused by secondary priming

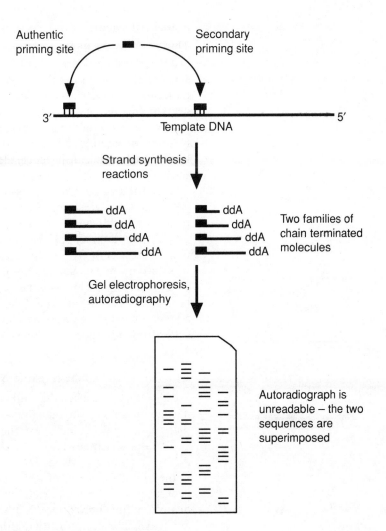

Fig 3.18

Our primer

Fig 3.18

Our primer

5′   3′
  AAGCTAGCTGAGTATCGTCC

◇ The 'melting temperature' or T$_m$ (equivalent to the maximum annealing temperature) of the primer can be approximated from the formula:
T$_m$ = (4 × number of Gs and Cs) + (2 × number of As and Ts).
In our example T$_m$ = (4 × 10) + (2 × 10) = 60°C. A single mismatch (i.e. position at which base pairing cannot occur) will reduce the T$_m$ by about 5°C.

sequences related to the desired priming site might be present, sequences that share, say, 12 of the 15 nucleotides of the main site. If the annealing step is not performed correctly then primers might anneal to these related sequences by mistake.

The critical factor is the temperature at which annealing is carried out. To illustrate this we will take as an example a primer which is 20 nucleotides in length and has a GC content of 50 per cent (*Figure 3.18*). This primer will be able to form a stable base paired structure with its target site on the template molecule at temperatures up to about 60°C in the Tris-MgCl$_2$ buffer used for annealing. At higher temperatures the base pairing is broken by thermal effects and the primer dissociates from the template. The important point is that a partially base paired structure is less able to withstand the thermal effects, and so primers annealed to secondary sites detach at lower temperatures. In our example, primers annealed to a secondary site at which only 19 base pairs can form will be unstable at temperatures above about 55°C. We could therefore ensure that priming is specific by carrying out the annealing step at 55–60°C.

◇ In practice you need worry about specificity and efficiency of priming only if you are using a primer directed at an internal site within your cloned DNA fragment. If you are using a universal primer then simply use the optimal annealing temperature, annealing time, and cooling regime recommended by the supplier.

Efficiency of priming

This refers to the number of template molecules that end up with a primer attached to them. Clearly, you want this number to be close to 100 per cent, so that you can generate as many chain terminated molecules as possible and have nice strong bands in your sequencing gel. On the other hand, you do not want to have too many primers, as any that are left over when you start to cool down your annealing mixture may be able to attach to secondary sites on the template, leading to non-specific priming. Ideally the molar ratio should be 1:1, so that you have one primer for every template molecule.

To be able to calculate the amounts of primer and template to use you must know their concentrations, which in the case of the template may be difficult to work out. Yields from single stranded DNA preparations are very variable and the amounts are generally too low to measure accurately. For this reason, most protocols err on the side of safety by recommending amounts of primer and template that result in a template excess if the yield has been high. Even if you have been less successful with your DNA preparation you should still have a small excess of template in your annealing mixture and will not suffer from non-specific priming.

◇ The annealed template-primers can be kept in the −20°C freezer for several months if you have to delay the rest of the experiment for any reason.

2.1.2 The strand synthesis reactions

We have finally reached the bit that you have been waiting for. In the next 20 minutes or so you will make the chain terminated molecules that are the central objective of the entire sequencing experiment.

The strand synthesis reactions are carried out in two steps (*Figure 3.19*). First you provide the DNA polymerase with limited quantities of the four dNTPs, one of which is labelled. The enzyme synthesizes new strands, but cannot extend them very far before running out of one of the four nucleotides. Importantly, the strands that are made are heavily labelled. Note that although the DNA polymerase stops making DNA the strands are not actually terminated, as the enzyme

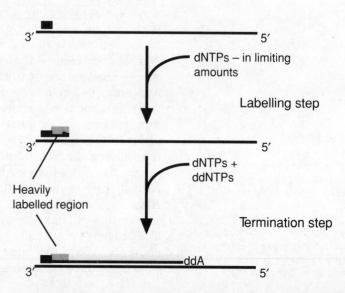

Fig 3.19

The labelling and termination steps

is still attached and can re-initiate synthesis in the second step of the procedure, when you add much larger quantities of the dNTPs together with the chain terminating ddNTPs. Now the polymerase extends the strands until a ddNTP is incorporated and a chain terminated molecule is produced.

Carrying out the strand synthesis reactions

◇ Basic safety procedures for handling radioactive material:
- Wear gloves
- Keep a Lucite or Perspex screen of suitable thickness between you and the radiochemical
- Ensure other lab personnel are not exposed
- Dispose of waste properly (i.e. in accordance with local regulations)
- Monitor yourself and your working area before, during and after all procedures involving radiochemicals

The first thing you must do is prepare the labelled nucleotide. If you are using radiolabelled material then you must, of course, follow the necessary safety procedures, which will be explained to you by your department or institute safety officer. It is a good idea to start by carrying out an entire sequencing experiment without the labelled nucleotide, so you can assure yourself (and your colleagues) that you can set up the reactions, run the sequencing gel, and dispose of the waste material without making a mess.

If the labelled nucleotide is in aqueous solution then remove it from the freezer it is stored in and place it behind a suitable screen. Do not at this stage open the safety container provided by the supplier. After about 30 minutes the solution will have thawed out and you can pipette the required amount of nucleotide. If the radionucleotide is dissolved in ethanol then it remains unfrozen at −20°C and you can pipette it as soon as you take it out of the freezer. However, ethanol interferes with the strand synthesis reactions, so you must repurify the nucleotide by evaporating the solution in a freeze drier. The dried nucleotides are then redissolved in one of the reaction components.

You must carry out four strand synthesis reactions in parallel, one for each nucleotide (*Figure 3.20*). To set up the labelling step you mix

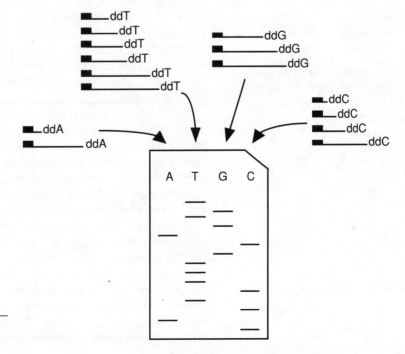

Fig 3.20

Four strand synthesis reactions are carried out in parallel

together the annealed primer-template, labelled and unlabelled dNTPs, and DNA polymerase in a Tris-NaCl-MgCl$_2$ buffer and incubate at room temperature for a short time, 10 minutes or so. The amounts of each reaction component that you use, as well as the incubation time, are important variables that you can change to suit the requirements of your experiment. We will discuss these factors in the next section.

At the end of the first incubation you initiate the termination reactions. All this requires is that you add additional unlabelled dNTPs, plus the appropriate ddNTP, into each reaction tube and incubate for a further 5 minutes or so. Again, you can vary the amounts of dNTP and ddNTP that you add. During this second incubation period the polymerase extends the strands until chain termination occurs. To complete the experiment you add a few microlitres of 'formamide dye mix', which contains formamide plus two dyes, xylene cyanol, and bromophenol blue. At this stage we do not need to worry about the role of the formamide dye mix—we will return to it when we load the polyacrylamide gel in Chapter 4, Section 1.2.3. If you have used a [^{32}P]dNTP then you should load and run the gel immediately, as the high emission energy of the label causes breakage of the chain terminated molecules (see Section 1.4). If you have used a [^{35}S]dNTP then you can relax as the chain terminated molecules can be stored in a −20°C freezer for up to a week. If you want to store the molecules for longer then leave out the dye mix.

Optimizing the variables

If you are using a sequencing kit then the supplier will have provided you with a detailed set of protocols for the sequencing reactions. These are designed to give good results but not necessarily the best you can hope for. By changing the absolute amounts and ratios of the dNTPs and ddNTPs, and by using longer or shorter incubation times, you can modify your experiment so that you obtain the clearest possible sequences from the regions of the template you are interested in.

The important point is that the strand synthesis reactions generate more sequence information than can be read from the polyacrylamide gel. A highly processive enzyme such as Sequenase can synthesize chain terminated strands up to about 3000 nucleotides in length. The strand synthesis reactions therefore have the potential to produce 3000 bp of sequence information. Unfortunately, at most only 500 bp of this sequence can be read from a single sequencing gel. You should therefore set up the sequencing reactions so that the bulk of the chain terminated molecules have lengths that correspond to the 500 bp region of the template that you are most interested in (*Figure 3.21*). Often this will be the 500 bp closest to the primer, but you may wish to extend a sequence you have already obtained by reading from positions 300 to 800, or from 500 to 1000. In fact you will be doing well to get beyond 800 bp on a single template, but this is not impossible and sometimes worth a try.

◇ One problem with some sequencing kits is that the dNTP and ddNTP solutions are pre-mixed at set concentrations, which means that you have little opportunity of changing the most important variable.

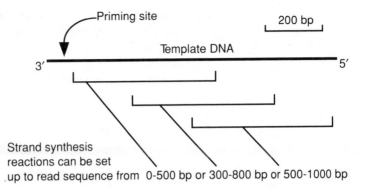

Fig 3.21

Which part of the template do you want
to sequence?

Strand synthesis
reactions can be set
up to read sequence from 0-500 bp or 300-800 bp or 500-1000 bp

To tailor your strand synthesis reactions in this way you need to
be in control of the variables. It is difficult to make hard and fast
rules here as your strategy depends on, among other things, the
concentration of your template solution, the GC content of your
template, the enzyme you are using, and the type of labelled nucleotide. The most important things to consider are as follows:

◇ Note that if your template has a high
GC content then you should decrease
the amounts of ddGTP and ddCTP that
you use, otherwise the chain terminated
molecules in the G and C families will, on
average, be shorter than the
corresponding A and T molecules.
Conversely, if the AT content is high,
reduce the ddATP and ddTTP
concentrations.

1. Chain termination does not occur during the labelling step, so this
   step can be used to set the lower lengths of the chain terminated
   molecules that are eventually obtained (*Figure 3.22*). If you include relatively large amounts of dNTPs in the labelling step, then
   you create a pool of relatively long molecules before any termination occurs. This would be appropriate for sequencing a region
   distant from the primer. On the other hand, by decreasing the
   dNTP amounts in the labelling step you can sequence close to the
   primer. You should be aware that the critical factor is not the
   absolute amount of dNTPs, but the ratio between dNTPs and
   annealed primer-templates. If your template is at a low concentration, or if primer annealing has not been efficient, then there are
   more dNTPs per extendible template, and the labelling step produces longer strands.

Large amounts of dNTPs in the labelling step

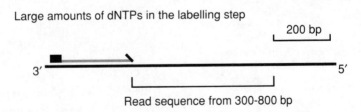

Read sequence from 300-800 bp

Small amounts of dNTPs in the labelling step

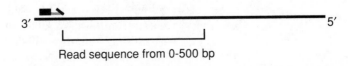

Fig 3.22

The amounts of dNTPs in the labelling
step can be used to set the region of the
template that you sequence

Read sequence from 0-500 bp

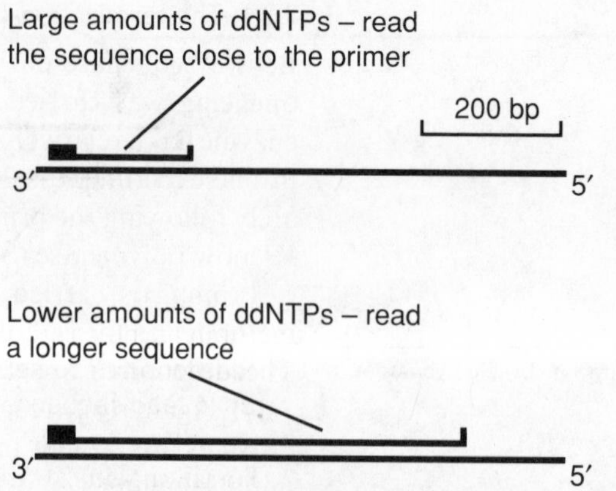

Large amounts of ddNTPs – read
the sequence close to the primer

200 bp

3′   5′

Lower amounts of ddNTPs – read
a longer sequence

3′   5′

Fig 3.23

The dNTP:ddNTP ratio determines the
length of sequence that is obtained

2. The time allowed for the labelling reaction is also important. If the incubation time is reduced then less strand synthesis occurs during the labelling step and shorter chain terminated molecules are obtained.

3. The dNTP:ddNTP ratio in the termination reactions determines how far strand synthesis proceeds, on average, before termination occurs. Greater amounts of ddNTPs give shorter sequences, lower amounts enable you to sequence further along the template (*Figure 3.23*). You must also take into account the relative affinity of the sequencing enzyme for dNTPs and ddNTPs as substrates. Sequenase, for example, has a four-fold preference for dNTPs compared to ddNTPs, so (in general terms) with an equal molar ratio of dNTPs and ddNTPs there is a 4:1 chance of a dNTP rather than ddNTP being incorporated at any nucleotide position. What this means is that if the template has a GC content of 50 per cent, a 1:1 dNTP:ddNTP ratio results in chain terminated strands that have an average length of 20 bp. If the molar ratio is 10:1 then the average lengths are increased to 200 bp, and at 40:1 to 800 bp. This assumes that you are using the standard Sequenase buffer, which contains Mg^{++} ions. If you replace the Mg^{++} with Mn^{++} then the enzyme's affinity for ddNTPs is increased so both dNTPs and ddNTPs are used equally efficiently (Tabor and Richardson 1989*b*). This reduces the average length of the chain terminated strands: a useful way of obtaining short sequences near to the primer.

◇ If the GC content is 50 per cent then there will be equal numbers of As, Cs, Gs and Ts in the template. Taking the 'A' reaction (containing ddATP) as an example, if the enzyme has a 4x preference for dATP compared to ddATP, then on average it will incorporate four dATPs into the growing strand for every ddATP, i.e. ...A..A...A...A...ddA. An A will occur on average once every four nucleotides, so the average length of the chain terminated molecules will be 20 bp.

◇ For a good discussion of the factors affecting the lengths of the chain terminated molecules see Slatko et al. (1993).

2.2   Alternatives to the labelling-termination procedure

The labelling-termination procedure is the best choice for sequencing a single stranded DNA template with a highly processive enzyme such as Sequenase. There are, however, other chain termination sequencing procedures that you may occasionally wish to use. We will look at three of these.

2.2.1 The termination-chase procedure with Klenow polymerase

Before Sequenase enzymes became available chain termination sequencing was carried out with Klenow polymerase. Because this enzyme has a relatively low processivity the best that can be hoped for in one experiment is 200–300 bp of sequence in the region immediately following the priming site. The sequencing procedure used with Klenow polymerase (Sanger *et al.* 1977) reflects this limitation.

Priming is carried out as described for the labelling-extension method, and the radiolabelled nucleotide is handled in the same way. The differences are seen in the strand synthesis reactions (*Figure 3.24*). Again, four reactions are set up in parallel, but in this case the ddNTPs are present from the beginning of the reaction. There is no initial labelling step with just dNTPs, labelling and chain termination occurring in the single reaction mixture. After the initial incubation additional dNTPs are added to 'chase' the reaction. The idea is that any strands that have not been terminated with a ddNTP are now extended as far as possible, certainly to lengths greater than 300 bp. All of the molecules resolved in the <300 bp region of the sequencing gel are therefore terminated with a ddNTP, rather than resulting from the enzyme running out of dNTPs and stalling. Stalled molecules would produce ghost bands on the polyacrylamide gel and make it difficult to read the sequence.

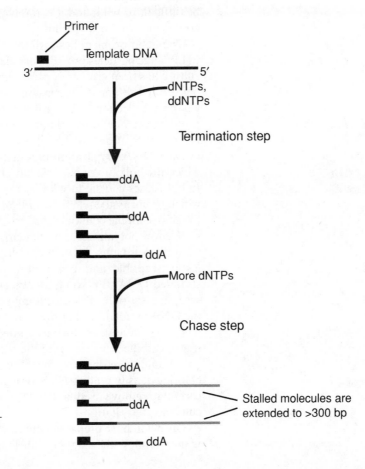

Fig 3.24

The strand synthesis reactions in the termination-chase procedure

With the termination-chase procedure there is little opportunity for customizing the reaction conditions to suit particular requirements. The method can provide 300 bp of sequence adjacent to the primer but little else. It is, however, extremely good at what it does and is often a better alternative than the labelling-extension procedure if you want to obtain a sequence that begins very close to the priming site.

2.2.2 Sequencing double stranded DNA templates

You will remember from Chapter 2 that quite complicated procedures are required in order to obtain single stranded DNA for chain termination sequencing. You may have asked yourself whether these procedures could be avoided by attempting to use double stranded DNA as the template. If we could sequence double stranded DNA then we would not need to clone our DNA into M13 or phagemid vectors, instead being able to use our friendly plasmid or bacteriophage λ systems.

◇ For a full description of double stranded sequencing see Murphy and Ward (1989).

You are not the first person to appreciate the potential advantages of double stranded DNA sequencing. Over the years there have been several attempts to develop efficient procedures, but even today the results do not bear comparison with the conventional single stranded methods. Sequences generated from double stranded DNA tend to be difficult to read, and it is rarely possible to be sure of more than 200 bp or so in one experiment.

The main problem lies with the quality of the DNA preparation. M13 phage particles provide very clean DNA because, if the infected culture has been grown properly, there is very little cellular contamination of the phage suspension used as the starting material (see Chapter 2, Section 1.4.1). All you have to do is remove the phage protein coat and you have pure template DNA. In contrast, plasmid DNA preparation involves breaking open the host bacterial cells, and a λ DNA preparation uses as the starting material a culture in which the bulk of the bacteria have been lysed by the bacteriophage. In both cases you are faced with a much more complex mixture from which to purify the template and you will find it much more difficult to remove all the contaminants.

Klenow polymerase is particularly sensitive to contamination and should never be used with a double stranded template. Sequenase is less susceptible and can produce short sequences from double stranded DNA. But you must still ensure that the template DNA is scrupulously pure. RNA contaminants, for instance, can anneal to the DNA template and prime strand synthesis reactions. By now you will understand the unfortunate affect this would have on your ability to read the sequence from the polyacrylamide gel. With plasmids you should therefore carry out a full scale purification with a caesium chloride density gradient step, or if you must use a mini-prep then be careful to remove residual RNA by ribonuclease digestion or lithium chloride precipitation.

Once you have prepared your ultrapure double stranded template you can follow the standard labelling-termination procedure to ob-

tain a sequence. The only modification is that you must fully denature the double stranded DNA before annealing the primer. This is
achieved by treating with alkali or by boiling. Both methods break
the base pairs between the two strands and leave single stranded
regions which hopefully include the priming site.

2.2.3 Thermal cycle sequencing

This is a recent innovation that makes use of the thermostability of
enzymes such as *Taq* DNA polymerase in a procedure that was
originally designed for sequencing DNA fragments produced by
PCR amplification. Obtaining a single stranded template for conventional chain termination sequencing of PCR products can be time-
consuming (Chapter 2, Section 2.2). Thermal cycle sequencing avoids
the problem by enabling you to sequence the PCR product directly.

The procedure works as follows. After the first PCR experiment,
you remove leftover primers from the amplified DNA by column
chromatography. You then use a portion of the amplified DNA as
the starting material for a second amplification, but this time you
add just one of the two PCR primers and include ddNTPs in the
reaction (*Figure 3.25*). During the thermal cycling the thermostable
polymerase treats one strand of the amplified DNA as a sequencing
template, and synthesizes chain terminated molecules that provide
the required banding pattern on the sequencing gel. You can include
a labelled nucleotide in the reaction, but better results are usually
obtained if the primer is end-labelled (Section 1.4.1). This is because
it is difficult to completely remove the primers from the first PCR, so
some chain terminated molecules representing the sequence of the
second strand are also produced. If your sequencing primer is end-
labelled then these unwanted molecules do not show up on the sequencing autoradiograph, and so can be ignored.

Thermal cycle sequencing works well if your initial PCR produces
a single product, so you see only one band when you analyse the

◇ Several companies market suitable
pre-packed chromatography columns.

Product of a standard PCR

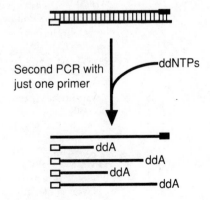

Second PCR with
just one primer

ddNTPs

ddA
ddA
ddA
ddA

Chain terminated strands –
numbers increase as more
cycles are carried out

Fig 3.25

Thermal cycle sequencing

results on an agarose gel. If your PCR produces additional bands then you must purify the desired PCR product from the agarose gel before carrying out the sequencing reactions. The thermal cycling procedure can also be used to sequence small amounts of double stranded plasmid and λ templates (e.g. from a single colony or plaque), but as in standard double stranded sequencing the DNA must be absolutely free from contaminants.

Further reading

Brown, T.A. (ed.) (1991). *Molecular biology labfax*. BIOS Scientific Publishers, Oxford—provides details of DNA polymerases used in chain termination sequencing, radionucleotides and modified nucleotides.

Slatko, B.E., Albright, L.M., and Tabor, S. (1993). DNA sequencing by the dideoxy method. In *Current protocols in molecular biology*, (ed. F.M.Ausubel *et al.*), pp. 7.4.1–7.4.27. Greene Publishing Associates and John Wiley and Sons, New York—an excellent description of chain termination sequencing protocols plus background information.

References

Barr, P., Thayer, R., Najarian, R., Seela, F., Laybourn, P., and Tolan, D. (1986). 7-deaza-2′-guanosine triphosphate: enhanced resolution in M13 dideoxy sequencing. *Biotechniques*, **4**, 428.

Biggin, M.D., Gibson, T.J., and Hong, G.F. (1983). Buffer gradient gels and ^{35}S label as an aid to rapid DNA sequence determination. *Proceedings of the National Academy of Sciences, USA*, **80**, 3963.

Innis, M., Myambo, K., Gelfand, D., and Brown, M. (1988). DNA sequencing with Thermus aquaticus DNA polymerase and direct sequencing of polymerase chain reaction-amplified DNA. *Proceedings of the National Academy of Sciences, USA*, **85**, 9436.

Martin, C., Bresnick, L., Juo, R.-R., Voyta, J.C., and Bronstzin, I. (1991). Improved chemiluminescence DNA sequencing. *Biotechniques*, **11**, 110.

Mills, D.R. and Kramer, F.R. (1979). Sequence independent nucleotide sequence analysis. *Proceedings of the National Academy of Sciences, USA*, **76**, 2232.

Murphy, G. and Ward E.S. (1989). Sequencing of double-stranded DNA. In *Nucleic acids sequencing: a practical approach*, (ed. C.J.Howe and E.S.Ward), pp. 99–115. IRL Press at Oxford University Press.

Prober, J. *et al.* (1987). A system for rapid DNA sequencing with fluorescent chain-terminating dideoxynucleotides. *Science*, **238**, 336.

Sanger, F., Nicklen, S., and Coulson, A.R. (1977). DNA sequencing with chain-terminating inhibitors. *Proceedings of the National Academy of Sciences, USA*, **74**, 5463.

Tabor, S. and Richardson, C.C. (1987*a*). Selective oxidation of the exonuclease domain of bacteriophage T7 DNA polymerase. *Journal of Biological Chemistry*, **262**, 15330.

Tabor, S. and Richardson, C.C. (1987*b*). DNA sequence analysis with a modified bacteriophage DNA polymerase. *Proceedings of the National Academy of Sciences, USA*, **84**, 4767.

Tabor, S. and Richardson, C.C. (1989*a*). Selective inactivation of the exonuclease activity of bacteriophage T7 DNA polymerase by in vitro mutagenesis. *Journal of Biological Chemistry*, **264**, 6447.

Tabor, S. and Richardson, C.C. (1989*b*). Effect of manganese ions on the incorporation of dideoxynucleotides by bacteriophage T7 DNA polymerase and *E. coli* DNA polymerase I. *Proceedings of the National Academy of Sciences, USA*, **86**, 4076.

4 Chain termination sequencing: running the gel and reading the sequence

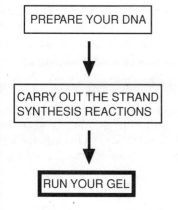

PREPARE YOUR DNA

↓

CARRY OUT THE STRAND
SYNTHESIS REACTIONS

↓

RUN YOUR GEL

The strand synthesis reactions provide you with four families of chain terminated molecules. In one family all the molecules terminate with A, in another they all end in C, in the third with G, and in the fourth with T. You must now separate these molecules in a polyacrylamide gel in order to obtain the banding pattern from which the sequence of the template DNA can be determined. In outline this involves:

1. Gel electrophoresis (Section 1): preparing a polyacrylamide gel, loading your samples, running the gel, and obtaining an autoradiograph (assuming you have used a radioactive label).

2. Reading the sequence (Section 2): interpreting the banding pattern on the autoradiograph.

3. Sequence analysis (Section 3): assembling individual sequences obtained from different experiments into a contiguous master sequence, identifying open reading frames and other genetic motifs, and searching databases for related sequences.

This chapter takes you through these three steps.

1. The sequencing gel

Gels for DNA sequencing tend to messy and temperamental. As you pour your first gel, which will almost certainly cause you a lot of problems, you may find it hard to believe that sequencing gels represent more or less the ultimate in current gel electrophoresis technology. For sequencing to be successful it must be possible to separate DNA molecules whose lengths differ by just a single nucleotide. This is an absolute requirement if the families of chain terminated molecules are to be resolved in a way that enables the sequence of the template to be determined. It is not easy to separate a DNA strand that is, say, 147 nucleotides long from one that is 146 or 148 nucleotides, but it can be achieved by ultrahigh resolution polyacrylamide gel electrophoresis. The next few pages tell you how.

1.1 What happens in gel electrophoresis?

DNA molecules, like proteins and many other biological compounds, carry an electric charge, negative in the case of DNA. This means that when DNA molecules are placed in an electric field they migrate towards the positive pole. This process is called electrophoresis.

Conventional electrophoresis is carried out in solution phase. Under these conditions the rate of migration of a molecule depends on two factors, its shape and its charge-to-mass ratio. Most DNA molecules are the same shape (linear) and have the same charge-to-mass ratio. As a result, DNA molecules of different lengths cannot be resolved effectively by conventional electrophoresis. The length of a DNA molecule does, however, become a factor if the electrophoresis is performed in an agarose or polyacrylamide gel. A gel comprises a complex network of pores through which the DNA molecules must travel to reach the positive electrode. The shorter the DNA molecule, the faster it can migrate through the pores of the gel. Gel electrophoresis therefore separates DNA molecules according to length (see *Figure 1.1*), with the resolution and the sizes of molecules that can be separated depending on the pore size of the gel. For sequencing you need a resolution of one nucleotide and must be able to separate molecules in the range 20 to 1000 nucleotides. This cannot be achieved with an agarose gel as the pores are too big, so you have to use polyacrylamide.

◇ The dependence on shape is the reason why hairpin loops in the chain terminated molecules are a problem. See Chapter 3, Section 1.3.

1.2 How to handle a polyacrylamide gel

Polyacrylamide gels are formed by the polymerization of acrylamide monomers into long chains cross-linked by N,N'-methylenebisacrylamide units (*Figure 4.1*). The pore size is determined by the concentration of the acrylamide monomers in the polymerizing solution (Andrews 1991). For DNA sequencing you generally use a 6 per cent gel, which resolves molecules in the range 25–500 nucleotides. If you wish to read a sequence very close to the primer then you should increase the gel concentration to 8 per cent, or if you aim to extend a sequence beyond 500 nucleotides then you should reduce it to 5 per cent or even 4 per cent.

1.2.1 Preparing a polyacrylamide gel

To prepare a polyacrylamide gel you must mix together the following components:

◇ Recipes for polyacrylamide gels are given in Ward and Howe (1991) and Slatko and Albright (1993)—see Further Reading.

- acrylamide
- N,N'-methylenebisacrylamide (generally referred to as 'bis')
- ammonium persulphate

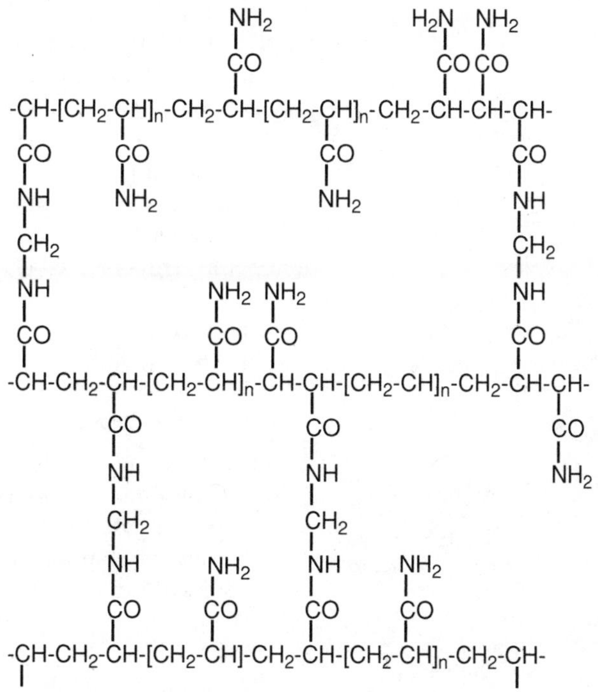

$$\begin{array}{c}
\text{NH}_2 \qquad\qquad\qquad\quad \text{H}_2\text{N} \quad \text{NH}_2 \\
| \qquad\qquad\qquad\qquad\qquad | \quad\; | \\
\text{CO} \qquad\qquad\qquad\qquad\;\; \text{CO}\,\text{CO} \\
| \qquad\qquad\qquad\qquad\qquad | \quad\; | \\
\text{-CH-[CH}_2\text{-CH]}_n\text{-CH}_2\text{-CH-[CH}_2\text{-CH]}_n\text{-CH}_2\text{-CH-CH-CH-} \\
\end{array}$$

Fig 4.1

A typical polyacrylamide gel matrix

- urea
- electrophoresis buffer
- N,N,N',N'-tetramethylethylenediamine (TEMED)

What are all these things for?

Acrylamide and bis

◇ Several companies supply ready prepared acrylamide-bis stock solutions. These save you from the dangers of handling the powders and are recommended if you prepare sequencing gels on a regular basis.

These are the monomeric units that form the gel matrix. Both are usually obtained as powders and the first thing you must do is prepare a concentrated stock solution from which you can later remove aliquots when you make the gel itself. The most convenient stock solution is one with a total monomer concentration of 40 per cent, comprising 38 per cent acrylamide and 2 per cent bis.

You must always take the utmost care when weighing out and dissolving the acrylamide and bis powders. Both are neurotoxins and can enter your body by inhalation or by absorption through the skin. The best policy is to minimize the extent to which you handle the powders by not weighing out the acrylamide when making your stock solution. Rather than making, say, 250 ml of the 40 per cent stock (which would mean weighing out 95 g of acrylamide and 5 g of bis) you should make whatever volume is appropriate for the amount of acrylamide in the unopened reagent bottle. For example, if you have a 100 g jar of acrylamide then empty it all into a 500 ml beaker (you should wear gloves, a face mask, and use a fume cupboard). Add 5.26 g of bis (keep the gloves and facemask on when you weigh this out) and then dissolve in 220 ml water. When all the powder is dissolved add more water to make the solution up to 263 ml. You

◇ If you are the slightest bit doubtful about the safety precautions for handling toxic powders then talk to your lab supervisor or safety officer.

now have 40 per cent (38:2) acrylamide-bis stock solution. The dissolved acrylamide is still dangerous, and you must always bear this in mind when handling the solution. Acrylamide remains toxic until polymerized, and even then should be treated with caution.

Some batches of acrylamide powder are contaminated with metal ions which interfere with the electrophoresis. To remove these you should add a small amount (up to 5 per cent w/v) of the mixed bed ion exchange resin called Amberlite MB-1, and stir gently for an hour at room temperature. The resin absorbs the metal ions and is then removed by filtering the solution. Store your stock solution in the fridge or cold room (don't forget the biohazard label). It will keep for about three months.

◇ You can check the freshness of an acrylamide stock solution by measuring its pH. If the pH is less than 7.0 then it should be OK to use. Higher pH values indicate that breakdown to acrylic acid has occurred, and the stock should be discarded.

Ammonium persulphate

This compound initiates the gel polymerization. You should make a fresh solution of ammonium persulphate each time you prepare the gel. You can add the ammonium persulphate at any time when you are mixing the components of the gel, as the initiated reaction cannot proceed without the catalyst called . . .

◇ In fact, you can store the ammonium persulphate solution in the fridge (+4°C) for several weeks. This is not the best policy, however, as the solution gradually becomes inactive, meaning that it takes longer for the gel to polymerize completely. You then run the risk of loading the gel too soon, leading to fuzzy bands (see Section 2.2.4).

TEMED

. . . which is usually the last thing you add to your mix, immediately before you pour the gel. TEMED is a liquid that is stored in the fridge or cold room in a dark bottle.

Urea

Urea is a denaturant and is included in the gel so that hairpin loops are less likely to form in the DNA molecules being electrophoresed. Remember that hairpin loops alter the mobility of a DNA molecule and interfere with the banding pattern (see Chapter 3, Section 1.3). Usually urea is added into the gel mix as a solid to give a final concentration of 7 M.

Formation of hairpin loops in the chain terminated molecules is one of the major difficulties encountered with DNA sequencing and the addition of urea is just one of several things aimed at alleviating this problem. We have already seen that you can replace the dGTP in your strand synthesis reactions with a modified nucleotide in order to reduce intrastrand base pairing (Chapter 3, Section 1.3). The fact that the polyacrylamide gel heats up to 65°C or so during electrophoresis also keeps base pairing to a minimum. As a last resort, the urea in the gel can be supplemented with a second chemical denaturant, such as formamide (Brown 1984).

Electrophoresis buffer

◇ 10x TBE = 89 mM Tris-borate (pH 8.3), 2 mM EDTA. There is no need to adjust the pH.

The final component in the gel mix is the electrophoresis buffer which provides the ionic solution needed to conduct electricity from one electrode to the other. Sequencing gels are run at high voltage so it is important that the buffer has a high capacity, otherwise it might become depleted during a run. You should use TBE buffer, at 1x strength in the final gel mix.

1.2.2 Preparing the gel plates and pouring the gel

Polyacrylamide gels are electrophoresed vertically with the gel enclosed between two glass plates. The exact dimensions of the plates depend on individual taste and on the size of the electrophoresis apparatus that is available. Typically, the plates are 40 cm long and 20 cm wide. This is long enough to achieve a good separation for 200–300 bands, and wide enough to accommodate 32–40 lanes. You are therefore able to electrophorese the chain terminated molecules from a number of sequencing experiments in one go.

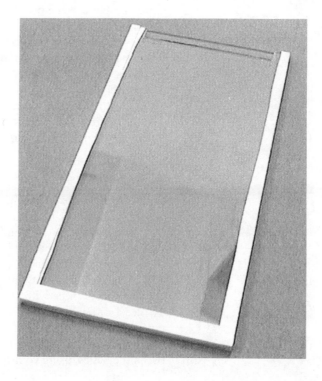

Fig 4.2

A typical plate assembly, taped up and ready for gel pouring

Assembling the plates

A typical plate assembly, ready for gel pouring, is shown in *Figure 4.2*. There are four things to note:

1. The two plates are not the same size. One plate is shorter than the other and may have 'ears' or 'notches' at either side. Among other things, this helps us when we pour the gel.

◇ Silanizing solution is toxic by inhalation: use it in a fume hood and wear gloves. With some protocols only one plate is silanized, so the gel adheres to the second plate when the assembly is taken apart after electrophoresis, and is prepared for autoradiography in a manner slightly different from that described in the text (Section 1.3.1).

2. Although not apparent in *Figure 4.2*, the inner surfaces of the two plates are totally dust-free, having been cleaned with water and rinsed in ethanol. Silanizing solution has also been applied to the two inner surfaces. This solution applies a silicone polish, which prevents the gel adhering to the plates when the assembly is taken apart after electrophoresis.

3. The two plates are held apart by plasticard spacers, 1 cm or so in width, laid down either side of the assembly. These spacers determine the thickness of the gel, which in turn has an influence on the sharpness of the bands that are eventually obtained. You will probably use gels 0.3 mm thick for most sequencing experiments, but bear in mind that this can be reduced to 0.2 mm if band resolution is a problem.

4. The plates are held together very tightly. Although there are some systems that use bulldog clips and a prayer, it is best to hold the assembly together with tape. You should use a heat resistant tape as the plates get up to about 65°C when you run the gel. You should be very careful to seal the plates together effectively, to prevent leakage when you pour the gel. Make sure that there are no wrinkles in the tape along the sides and the bottom of the plates, and arrange the tape very neatly around the corners. If you tape the plates properly then leakage of the gel is not a problem. Tricks to reduce leakage, such as sealing the edges of the assembly with agarose, are necessary only if you use substandard tape or cannot be bothered to tape up carefully.

◇ To prevent leaks you need to achieve the level of neatness appropriate to the wrapping of a birthday present for a greatly loved grandparent.

Pouring the gel

In one hand you have your gel mix, in the other your taped up plate assembly. You have just added TEMED to the gel mix and know that in 5–10 minutes the acrylamide will have polymerized. The question that you are asking yourself is: how on earth do I get the gel mix into the 0.3 mm gap between the plates without making loads of bubbles?

This is not easy but once you have mastered it you have an excellent party piece with which to impress new graduate students. The least difficult method is to draw the gel mix into a 25 ml glass pipette, remove the suction device, and hold the mix in the pipette with your finger. Then gradually release the mix into the space between the plates, using your finger to control the flow. Three tips:

◇ Pouring a polyacrylamide gel is one of those techniques that is best learnt by watching someone who knows how to do it.

- Hold the plates at just the right angle, so the gel mix flows down one side and gradually fills the space between the plates (*Figure 4.3*)
- Keep the flow of the gel mix as even as possible; bubbles form when the flow speeds up or becomes interrupted
- Don't panic unless you really want to

As soon as the space between the plates is filled, lay the assembly down with the open end raised slightly by placing it on a 10 ml glass pipette. Slide the comb into the open end. This may be a conventional comb or the sharkstooth variety, but we will worry about that in Section 1.2.3 when we load the gel. Attach bulldog clips to the top of the assembly to hold the comb tightly in place and then leave the whole thing undisturbed for at least 30 minutes (possibly more, depending on the gel recipe you have used) to allow the acrylamide to polymerize completely. You can keep the gel overnight at room

◇ If you are using a conventional comb then do not push it in too far. The top of the wells should be in line with the top of the shorter plate, as shown below.

Back plate   Front plate

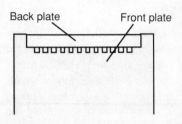

Fig 4.3

Pouring a gel

temperature, but make sure the open end stays moist by covering this part of the assembly with tissues soaked in electrophoresis buffer.

Dealing with spills, leaks, and bubbles

A certain amount of spillage is to be expected when you pour the gel. This is nothing to worry about as polymerized gel mix can be treated as low hazardous waste. Leave the spillages for 30 minutes or so to polymerize and then clean up with paper towels, water, and detergent.

There is not much you can do about leakages from between the plates. If you have a major leak then you must give up and put together a new plate assembly. If the leak is less major, then finish pouring the gel as normal. When you lay the plate assembly flat the leak will probably ease up, and if you are lucky the gel will polymerize before too much of the mix has dripped out. A small empty area in one corner does not affect the electrophoresis unduly.

Bubbles that get into the gel need to be removed if at all possible or at least eased to one side so that the maximum number of lanes are clear. Over-filling is often the best way to force bubbles out of the top of the assembly. Alternatively, cut a thin piece of plasticard (the same type as used to make the spacers), slide this between the plates, and attempt to move the bubbles to wherever you want them. Do not play around like this for more than 15 seconds. Once the gel starts to set you do not want to disturb it too much as you risk disrupting the

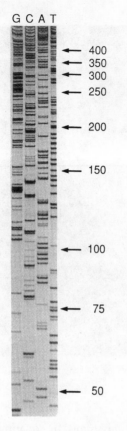

G C A T

← 400
← 350
← 300
← 250

← 200

← 150

← 100

← 75

← 50

Fig 4.4

A sequencing autoradiograph showing bunching of bands towards the top of the gel. (Reproduced with permission from Slatko *et al.* (1993), page 7.4.21. ©1993).

◇ Some assemblies include a metal backing plate that is clamped on to the glass plates. The metal plate helps ensure an even heat distribution across the gel. If the sides of the gel are cooler than the middle then the outer lanes will run more slowly, leading to a 'smiling' effect.

polymerization process. This may lead to discontinuities within the gel, with a detrimental effect on the banding clarity.

Advanced gel pouring

One problem with standard polyacrylamide gels is that the band separation is not even throughout the entire length of the gel. The bands that move the furthest become more spaced out, whereas the slower moving ones bunch together. When you attempt to read the sequence you find that working out the band order becomes difficult at the 'top' of the gel as the bands are too close together to separate by eye (*Figure 4.4*).

There are a number of ways of producing a more even banding pattern, the two most popular strategies being:

1. Wedge gels (Olsson *et al.* 1984), which are thicker at the bottom than at the top. This creates an electrical field gradient within the gel, decreasing as the gel becomes thicker. The migration rates of the DNA molecules therefore decrease as they progress towards the bottom of the gel, producing a more even banding pattern. A wedge gel is produced with wedge shaped spacers. Wedge gels are no more difficult than ordinary gels to pour, but are more tricky to handle when you take the assembly apart at the end of the electrophoresis.

2. A buffer gradient gel (Biggin *et al.* 1983) has a variable concentration of TBE buffer, from 0.5x at the top to 5x at the bottom. Again, this creates an electrical field gradient, so the migration rate decreases as the molecules approach the bottom of the gel. Buffer gradient gels are more difficult to prepare than ordinary gels, as you have to combine two gel mixes, with different buffer concentrations, as you pour the gel.

1.2.3 Loading and running the gel

When the gel is fully polymerized you can attach the plate assembly to the electrophoresis apparatus. Polyacrylamide gels are run in a vertical position, with the wells at the top, so the DNA molecules migrate down through the gel. The electrophoresis apparatus has two buffer reservoirs, top and bottom, which you fill after the plates have been clamped in position. A typical set-up is shown in *Figure 4.5*.

Before placing the assembly in the electrophoresis apparatus you should slit the tape at the bottom of the gel with a clean razor blade. This enables an electrical contact to be made between the gel and the buffer in the lower reservoir. The comb is best left in the top of the gel until you have clamped the plates into the apparatus and filled the top reservoir. Then slide the comb out, being careful not to damage the top of the gel.

Conventional and sharkstooth combs

There two different types of comb. If you have used a conventional comb then you have made square cut-out wells in the top of the gel (*Figure 4.6A*). You must flush these out with buffer as soon as you

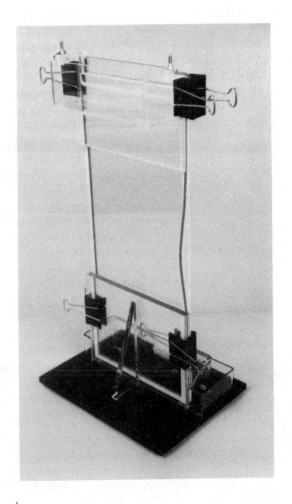

Fig 4.5

A simple universal length vertical electrophoresis system. A basic design 20 cm high electrophoresis apparatus has been sawn in two and the lower electrode cable extended. The two halves of the system are clamped to a standard glass plate assembly. The assembly supports the upper tank and can be of any length. Gel electrophoresis should be carried out with an appropriately sized safety cover on (not shown) or in a purpose built safety cabinet.

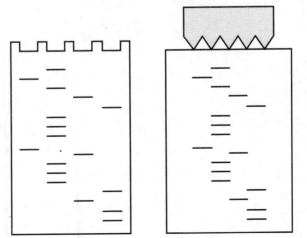

Fig 4.6

Conventional and sharkstooth combs

◇ Use a Pasteur pipette or syringe to flush out the wells.

remove the comb from the gel, and flush again immediately before loading the samples. The aim of this is to rinse out urea, which leaches from the top of the gel. Try to avoid squirting bubbles into the wells as bubbles lead to streaks when the gel is run.

A sharkstooth comb (*Figure 4.6B*) works in a rather different fashion. This type of comb does not make cut-out wells in the gel, but instead forms spaces into which you load your samples. You must remove the comb and flush the top of the gel, as just described, but then re-insert the sharkstooth comb and load your samples between its teeth. The advantage with a sharkstooth comb is that the lanes that are formed are absolutely adjacent, without the small gaps between them that you get with a conventional comb. This makes it easier to work out the band order near the top of the gel, but for many of us the benefit of this is outweighed by the fact that it is more difficult to load the samples with a sharkstooth comb.

Loading your samples

When we left your strand synthesis reactions in Chapter 3 you had just added the formamide dye mix to each sample. Before loading you must incubate the reactions at 95°C for 2 minutes to detach the chain terminated molecules from the template DNA. Then place the samples on ice and carefully load 2–4 µl of each reaction into the wells.

This may not be as easy as it sounds. The gap between the plates is less than half a millimetre so a normal pipette tip does not fit into the wells. You have four choices (*Figure 4.7*):

- use a special pipette tip, with a drawn out end, which any one of several companies are happy to sell to you
- use a drawn out capillary tube (other companies sell you suction devices to attach to these)
- use a 30 gauge or narrower syringe needle
- use an ordinary pipette tip, and let the sample drop vertically into the well

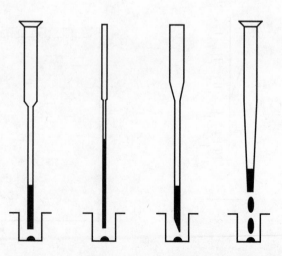

Fig 4.7

Different ways of loading your samples

The last method is the best one as it minimizes your chances of accidentally damaging a well during loading. You should not rush the loading but try to get finished in a couple of minutes so there is not time for any of the samples to diffuse out of the wells.

The order in which you load your samples is not important, as long as you remember what it is. Alphabetical order (A-C-G-T) is commonest but you may want to vary this. In particular, if your template DNA is highly AT or GC-rich then the order A-T-G-C is better, as this means that the two tracks with the majority of the bands (A and T, or G and C) are next to each other.

Running the gel

◇ A typical voltage would be 45 V cm^{-1}.

As soon as you have loaded the last sample, attach the leads to connect the apparatus to the power supply and switch on. The settings you use depend on the dimensions of the gel. Note that all sequencing gels require voltages that are high enough to kill you. Never tamper with the apparatus once it is switched on, and do not use an apparatus with a leaky reservoir.

◇ Some protocols recommend warming the gel up prior to loading by pre-electrophoresing, at the voltage setting to be used for electrophoresis, for 30 minutes or so.

You can follow the progress of the electrophoresis by watching the migration of the two dyes—bromophenol blue and xylene cyanol—down through the gel. The rates at which these dyes move depends on the concentration of acrylamide in the gel. In a 6 per cent gel the bromophenol blue dye (the faster one) moves at the same rate as a single stranded DNA molecule 25 nucleotides in length. Xylene cyanol in this gel moves at the rate of a 100 nucleotide molecule. You can therefore judge the length of run that is appropriate for the region of the sequence that you are interested in. If you want to read near to the primer then stop the electrophoresis when the bromophenol blue is a few cm from the bottom of the gel, which usually takes a bit less than 2 hours. If you wish to read a sequence away from the primer then run the gel for longer. If you have spare wells you can stop the electrophoresis at an appropriate point, load a second aliquot from each of your samples, and switch on again. This gives you a short and long run of the same sequence (*Figure 4.8*).

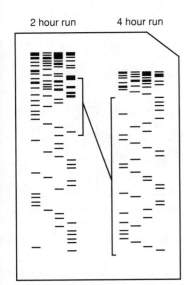

2 hour run   4 hour run

Fig 4.8

An autoradiograph showing a short and long run of a sequence

1.3 Autoradiography

After electrophoresis the banding pattern in the gel is visualized by autoradiography. This is the process by which β particles emitted by a radioactive atom such as ^{35}S or ^{32}P create an image on an X-ray film. The image is developed using standard photographic techniques (Mundy *et al.* 1991).

1.3.1 Preparing the gel for autoradiography

β particles are attenuated very quickly by glass, so we cannot simply lay the X-ray film on the surface of the plate assembly. Instead we must take the plates apart and transfer the gel to a piece of filter paper.

The first thing you must do is switch off the power supply and disconnect the leads. Then empty the buffer from the top and bottom reservoirs, remembering that the bottom buffer now contains radioactive material (short chain-terminated strands and unincorporated radionucleotides that have run right through the gel) and must be disposed of in an appropriate way. Remember also that parts of the gel itself are now radioactive, so any pieces that drop off must be checked for radioactivity before being thrown away.

◇ It is important that you do not lose track of the orientation of the gel (i.e. top and bottom, front and back). The best system is to cut off one corner of the gel, top right for example. The same thing applies with the X-ray film.

You should be able to prise the plates apart fairly easily. First you must remove just the shorter of the two plates, leaving the gel supported on the backing plate. This makes it easier to handle during the next step, which involves soaking ('fixing') in acetic acid to remove the urea. If this is not done then the urea crystallizes when the the gel is dried, causing a terrible mess. The fixing step takes about 15 minutes, depending on the thickness of the gel. Cover the gel with a piece of plastic mesh to stop it floating away in the fixer.

When the fix is complete, lift the plate-gel-mesh out of the acetic acid and drain it carefully. Hold the gel on to the plate with the mesh and try not to let the gel get crumpled up. If this does happen then put the plate on a flat surface and use a pair of blunt forceps to manipulate the gel back into shape. Squirting water between the gel and support plate helps here.

The next step is to remove excess liquid by blotting with paper towels, and then to gently press a piece of filter paper on to the gel. The gel adheres to the filter paper and you should be able to lift it off the plate without too much trouble. The gel is then dried on a vacuum drier at 80°C for 20–30 minutes. This step removes most of the moisture from the gel leaving a very thin film on the filter paper. If you used a ^{32}P label you could set up the autoradiograph with the undried gel, but the bands would be faint and fuzzy due to the β particles being scattered as they pass through the hydrated gel *en route* to the X-ray film. Much sharper bands, essential if you want to read a long sequence, are obtained from a dried gel.

◇ If you have used a ^{35}S label then you have no choice, you must dry the gel. ^{35}S has such a low emission energy that exposure of a wet gel takes for ever.

1.3.2 The autoradiograph

When the gel is dry you can set up the autoradiograph. This must be done in a dark room under a red safety light. Tape a piece of X-ray film on to the surface of the gel and put the whole thing in a light-tight cassette. The cassette is then left at room temperature for 24–48 hours before being developed.

This step sounds easy but poor practice can ruin your results. X-ray films are very sensitive things and rough handling leads to sharp and intense artefacts on the developed autoradiograph. Static electricity is a big problem, especially in hot weather, and you should avoid wearing nylon or 'plastic' gloves or overalls when handling the film; wear cotton gloves or none at all. Slide the film out of its box slowly, and do not bend it. Use the minimum amount of tape needed to hold the film in place, and remove this slowly after the exposure.

◇ With a Southern blot you might place the cassette at −70°C. This reduces background but only with long exposures (>1 week) so is not recommended for sequencing gels. You do not 'flash' the film.

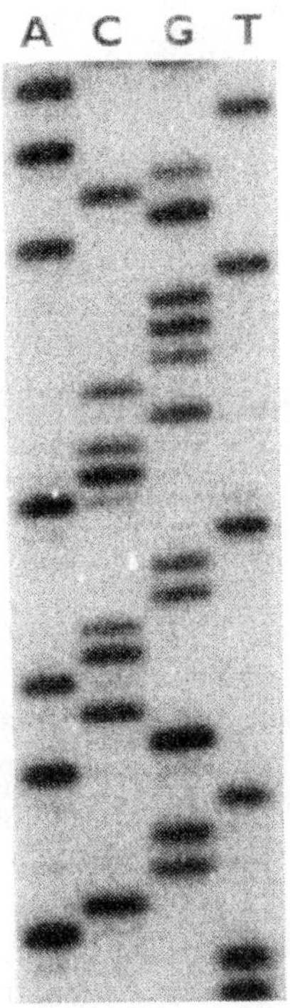

Fig 4.9

A cause for celebration. This is the sequence of part of a synthetic gene coding for an ice nucleating peptide. (Autoradiograph provided by Dr Anna V. Hine)

◇ When you are reading the autoradiograph, take account of the vertical spacing between the bands. Except in problem regions (see Section 2.3.1) the bands should be evenly spaced, though gradually getting closer together as you move towards the top of the gel. You can use the spacing to identify positions where a band is missing and to estimate the number of bands in a homopolymeric run (e.g. AAAAA), where the individual bands may merge together.

2. Reading the sequence from the autoradiograph

From the time that you started your single stranded DNA preparation we have been working in the dark with no reliable way of knowing if your experiment has progressed according to plan. Only when you develop the autoradiograph can you tell whether it is celebration time or back to the drawing board.

2.1 Celebration time: it has worked

We will start on an optimistic note and assume you have a beautiful clear banding pattern on your autoradiograph. If this is the case then reading the sequence is easy. First locate the band that has moved the furthest, the one that is closest to the bottom of the autoradiograph. This band represents the shortest chain terminated molecule. Note the lane that this band lies in, the T lane in the example shown in *Figure 4.9*. The first nucleotide in the sequence is therefore T.

The next most mobile band corresponds to a chain terminated molecule one nucleotide longer. Note the lane, T in our example. Sequence so far, TT.

Continue up through the autoradiograph, at each step identifying the next band in the series and noting its lane. In *Figure 4.9* the next four bands are in lanes A, C, G, G, so the sequence from the start is TTACGG.

2.2 Back to the drawing board: it has not worked

Now we must be pessimistic. What if the autoradiograph does not have a nice clear banding pattern? There are different degrees of failure but the most common problems are as follows:

- a completely blank autoradiograph
- an autoradiograph with some bands, but with other nasty marks, such as streaks and spots
- bands at all positions in one or more lanes
- the expected banding pattern, but faint or fuzzy
- ghost bands

Whichever category your autoradiograph falls into, you can use its appearance to deduce the most likely cause of your problems.

2.2.1 Nothing is visible at all

A total blank is unusual. Did you put the enzyme into the strand synthesis reaction? Was it the correct enzyme? Other than a technical error of this kind the only real possibility is that you did not have any bacteriophage particles when you started your single stranded DNA preparation. Check the single stranded DNA on an agarose gel (see Chapter 2, Section 1.4.1).

◇ For a complete treatment of problems encountered with sequencing gels, and the likely causes, see Slatko *et al.* (1993).

2.2.2 Streaks, spots, and other patterns

These effects generally point to a problem with the gel. If there is a dark streak in one or more lanes (*Figure 4.10A*) then you probably damaged the wells when you removed the comb, possibly because the gel was not completely polymerized. A thin streak within a lane, possibly with wavy bands (*Figure 4.10B*), indicates a bubble in the well or in the gel. Dark specks with trailing smears (*Figure 4.10C*) are caused by dust on the inside surface of the plates. Other more artistic patterns are due to static electricity generated by rough handling of the X-ray film.

2.2.3 Bands in all positions

The pattern shown in *Figure 4.11* indicates that one or more of the components of your strand synthesis reactions is in bad shape. As well as the molecules terminated by incorporation of a dideoxynucleotide, you have other molecules, of all possible lengths, terminated by some other process. This usually means that the polymerase has been unable to complete the strand synthesis reactions. The most likely culprit is the enzyme itself, which may have passed its use-by date or become partially inactivated by some other means. A shortage of one of the dNTPs in the strand synthesis reactions, through degradation or incorrect pipetting, is the next most likely cause. If just one lane is affected then you have probably used a dirty microfuge tube or pipette tip at some stage in the proceedings.

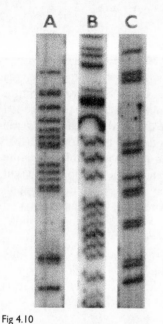

Fig 4.10

Problems with the gel: a damaged well (A), a bubble (B), dust (C)

2.2.4 Faint and/or fuzzy bands

This category encompasses a multitude of sins, too many to run through in detail. The exact nature of the faintness and fuzziness, and whether it affects all the gel or just specific parts, can be informative. If all the bands are faint then the yield of your single stranded DNA preparation was probably low. Check the preparation on an agarose gel and if necessary increase the amount of template in the strand synthesis reactions, although this may cause other problems by introducing more contaminants into the reactions. If your template yield appears to be good then poor annealing, too little primer, or not enough labelled dNTP may be to blame.

Changes in band intensity from top to bottom, in one or more lanes, indicate that the dNTP:ddNTP ratio needs to be adjusted. Faint bands at the bottom of the autoradiograph show that not enough termination has occurred for the length of sequence you are aiming for, so increase the ddNTPs in the relevant reactions. Faint bands at the top tell you to decrease the ddNTPs.

If all the bands are fuzzy then you must check your gel technique. The problem could be that the gel was not completely polymerized when you started the electrophoresis, the gel reagents were of low quality, you loaded too much sample into the wells, or you used the wrong power settings and overheated the gel. All are easily put right next time.

Fig 4.11

Bands in all positions

2.2.5 Ghost bands

◇ Ghost bands are common when dITP is used in the strand synthesis reactions.

Ghost bands are extra bands that appear at random in any of the four lanes. Although they are usually weaker than the main bands they complicate the banding pattern, often making it impossible to read the sequence. Ghost bands generally indicate that two or more sequences are being generated at once, which means that priming is occurring at secondary positions within your template DNA.

◇ If your template DNA is contaminated with *E. coli* genomic DNA then there is a high chance that secondary priming sites are present. Under these circumstances it may be impossible to solve the problem by changing the annealing conditions.

◇ If deletions are a problem then consider changing to a phagemid vector (see Chapter 2, Section 2.1).

Secondary priming may be due to your template DNA molecules having additional sequences similar to the target site (*Figure 4.12A*), in which case an increase in the annealing temperature usually solves the problem (Chapter 3, Section 2.1.1). If this does not work then the underlying cause of the ghost bands is more sinister. You may accidentally have taken two overlapping plaques when you infected the *E. coli* culture from which you made single stranded DNA. Alternatively deletions may have occurred in the insert DNA whilst the culture was growing (see Chapter 2, Section 2). In both cases the result is a template preparation that contains two or more different insert DNA sequences, each one preceded by an authentic priming site (*Figure 4.12B*). Result, ghost bands. The only solution is to go back to the beginning and make a new batch of single stranded DNA.

A. Secondary priming within the insert DNA

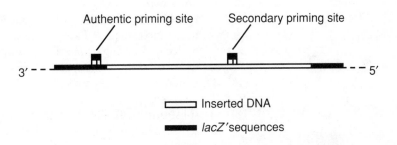

B. Template preparation contains two or more non-identical insert DNAs

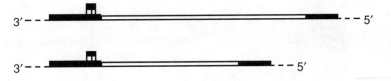

Fig 4.12

Two possible causes of ghost bands

2.3 Ambiguities within a clear sequence

If you are able to read your sequence with absolutely no difficulty then you really have done well. Even when you have built up a wealth of sequencing experience there will still be occasions when your banding pattern is far from ideal. This is because problems can arise not only as a result of technical shortcomings but also because of the nature of the sequence itself.

These sequence-dependent problems fall into two classes:

- abnormal banding patterns caused by hairpin loops in the template DNA or chain terminated molecules
- artefacts that are caused by the properties of the DNA polymerase

2.3.1 Problems with hairpin loops

We have already seen how base pairing in the template DNA or in the chain terminated molecules can produce hairpin loops (Chapter 3, Sections 1.2.3 and 1.3). Now we must examine the effect that these hairpin loops have on the banding pattern.

Compressions

Compressions are the most common artefacts seen on sequencing autoradiographs. A compression is a region where the vertical spacing between adjacent bands suddenly decreases, so that a series of bands run close together, possibly with two or more occupying the same position (*Figure 4.13*). Reading the sequence through a compression is usually impossible as you cannot be sure how many bands are superimposed.

Compressions occur when hairpin loops form in the chain terminated molecules. A hairpin results in a change in the shape of the molecule, altering its mobility within the gel. Molecules with hairpins tend to be more compact, so they migrate faster, resulting in the bands bunching together. If you can prevent the hairpins from forming then you will resolve the compression. The standard cures are substitution of dITP or 7-deaza-dGTP for dGTP in the strand synthesis reactions (Chapter 3, Section 1.3), addition of an extra denaturant such as formamide to the gel mix, or running the gel at a higher voltage so it heats up more (Section 1.2.3).

An alternative solution is to sequence the other strand of the template. To do this you need a clone in which the inserted DNA is present in the opposite orientation, so that when you prepare single stranded DNA you end up with the 'minus' rather than 'plus' strand of the template (*Figure 4.14*). When you sequence this new template you may find that the compression has shifted its position slightly, so that by comparing the sequences of the plus and minus strands you can work out the correct band order.

Hairpin loops in the template

Hairpin loops in the template DNA have a variety of effects on the banding pattern. The most common outcome is a discontinuity in band intensity, bands below a certain position in the autoradiograph being intense, and those above being weak. The drop in band intensity occurs because the polymerase has trouble breaking open the hairpin loop, so relatively few strands are synthesized into this region.

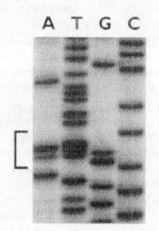

Fig 4.13

A compression. The correct sequence is within the bracket is CGTGATTTC

◇ Modified nucleotides are probably the first thing to try (dITP in preference to 7-deaza-dGTP). Running the gel at a higher temperature is less desirable as it increases the risk of a plate cracking.

◇ In fact you should always sequence both template strands in their entirety in order to confirm the sequence.

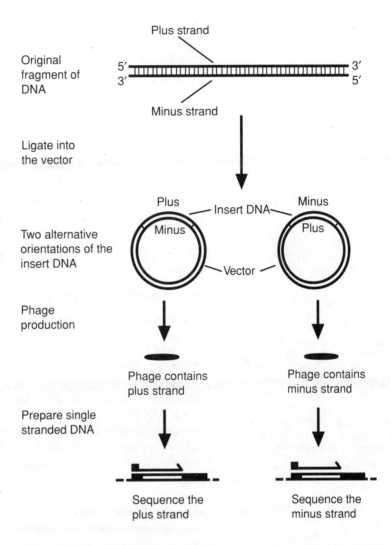

Fig 4.14

Sequencing the plus and minus strands of a template

◇ Walls are also caused by poor quality enzymes, and by contamination of template DNA with salt.

The junction between the intense and weak bands may be marked by a wall or pile-up (strong bands in all four lanes), due to the polymerase stalling and non-specifically terminating strand synthesis at the start of the hairpin.

The usual answer to hairpins in the template is to increase the temperature at which the strand synthesis reactions are carried out. This may mean changing to a thermostable DNA polymerase (Chapter 3, Section 1.2.4). A second possibility is to add into the strand synthesis reactions a non-enzymatic protein that binds to single stranded DNA, preventing hairpins from forming. Single-stranded binding protein (SSB) from *E. coli* can be used for this purpose. The DNA polymerase is able to strip off the SSBs as it moves along the template, so they do not interfere with strand synthesis.

2.3.2 Polymerase artefacts

Artefacts caused by the polymerase include faint bands and bands in unusual positions. These are fairly common but have the virtue of

Table 4.1 Polymerase artefacts (Ward and Howe 1989; Slatko *et al.* 1993).

Sequence feature	Artefact
Run of As	In a run of As the bottom band may be stronger than the others
Run of Cs	In a run of Cs the bottom band may be weaker than the others, and the next band up the strongest
TG	G is weak in the sequence TG
TGG	Second G is very strong
TGCC	Artefact band in the C track at or between the levels of the authentic T and G bands
GCA	Artefact band in the T or C tracks at the level of the authentic A band

Note. These artefacts are more common with Klenow polymerase.

◇ Artefacts may also be caused by pyrophosphorolysis, the removal of terminal ddNTPs as a result of attack by the pyrophosphate molecules generated during strand synthesis. Once the ddNTP is removed the polymerase can re-attach and resume strand synthesis. ddNTP removal is sequence-dependent, so only a few bands are affected, but those that are may be very weak on the autoradiograph.

occurring in a sequence-specific fashion. This means that they can usually be recognized for what they are and the correct sequence assigned to the region in question. The most common artefacts are listed in *Table 4.1*.

These artefacts arise in various ways. The faint bands are due to the influence that the sequence has on the affinity of the polymerase for a ddNTP. If this affinity decreases then less chain termination occurs resulting in a relatively weak band. Conversely, an increase in affinity for a ddNTP results in a more intense band. The reasons for these changes are not understood.

3. Sequence analysis

Your sequencing experiment has not reached its conclusion just because you have read the banding pattern on the autoradiograph. There are a variety of things still to do:

◇ The master sequence is called a 'contig', the short form of 'contiguous sequence'.

- if you are assembling a long master sequence, then you must establish if the new sequence overlaps with ones you have already obtained
- you may wish to search your sequence for restriction sites, repeated sequences, open reading frames, or other features
- you may wish to compare your sequence with other known sequences
- you should deposit your sequence with one of the three international DNA Databases

You can carry out a certain amount of sequence analysis by eye, using a pencil and paper. Generally, however, you need to enter your sequence into a computer and use a software package to perform the analyses. A number of commercial and home-grown packages are available for use on Macintosh or IBM personal computers (Cherry

1992), and most are very user-friendly, so don't worry if computers are not your strong point.

3.1 Entering your data into the computer

All software packages for DNA sequence analysis have an editor, which means that all you have to do to enter your sequence is type it in from the keyboard. This is a very simple process but of course you must not make any mistakes. Some packages have a voice synthesizer that 'talks' the sequence back to you as you press the keys. This alerts you if you accidentally hit the wrong key, but you should still double check when the whole sequence is entered.

More sophisticated systems have an automatic or semi-automatic data entry device of one type or another. The most reliable of these use a digitizer. You place your autoradiograph on a light box and then touch an electronic pen onto each band in turn. The digitizer converts this positional information into the sequence and enters the data directly into the computer. There are also one or two totally automated systems based on image analysers that scan the autoradiograph and 'read' the banding pattern, but in practice these devices are unreliable as they cannot make rational decisions about ambiguous bands, and may get caught out by background marks. In the end you must check the sequence by eye, so the automatic entry does not save much time.

3.2 Building up a contiguous sequence

Few interesting pieces of DNA are short enough to be sequenced in one experiment. If you are building up a long master sequence then you need a means of detecting overlaps so you can join new sequences on to the existing information (*Figure 4.15*). Often this can be done

Existing sequence is in two sections

    Section 1: TAGTCTTTCTAGAATCTA

    Section 2: ATGCTGATGCTGTATCGTT

New sequence contains an overlap . . .

New sequence
CGTTGCTGATGCTTAGTCT

ATGCTGATGCTGTATCGTT   TAGTCTTTCTAGAATCTA
     Section 2   Section 1

Fig 4.15

Detecting overlaps and building up a master sequence

. . . enabling a longer master sequence to be built

ATGCTGATGCTGTATCGTTGCTGATGCTTAGTCTTTCTAGAATCTA

by eye, but if you are using a random cloning strategy with a piece of DNA more than 3 kb in length then it may take time to find the desired overlaps. Most software packages can compare sequences and join overlapping ones together. The master sequence is therefore built up for you. The cleverer programmes warn you about disagreements between individual sequences, pointing out regions that need to be checked.

3.3 Searching for interesting features

Computer programs are particularly good at searching DNA sequences for features such as restriction sites, repeated sequences and open reading frames (ORFs). Some packages can also find potential tRNA genes and regions that might form hairpin loops in single stranded DNA or RNA.

The most useful of these applications is the ORF search as this enables you to identify potential protein coding regions (*Figure 4.16*). An ORF is a 'long' stretch of codons that begins with a initiation triplet (usually ATG) and runs for some distance before a termination codon (TAA,TAG, and TGA in most genomes) is reached. The word 'long' means 'however long the gene you are looking for might be'. If you are working with a cDNA, or bacterial genomic DNA, then ORFs that are protein coding regions are usually very easy to identify, as they are the only ones that stretch for anything approaching a reasonable length. The only complication is that each DNA sequence has six possible reading frames, three forwards and three

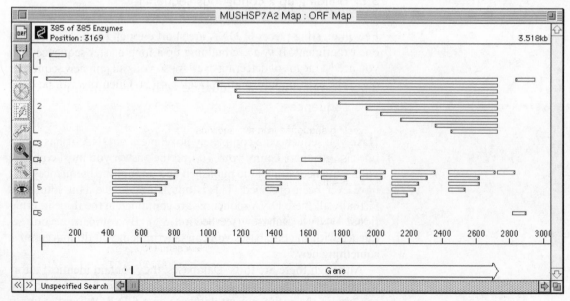

Fig 4.16

The results of an ORF search. A 3000-bp sequence has been searched by the computer in each of six reading frames (numbers at the left of the window). Each rectangle is an ORF that begins with an ATG initiation codon. The gene lies in the second reading frame, running from position 750 to 2700. The shorter 'ORFs' in this region of reading frame 2 indicate the positions of ATG codons within the gene, which the computer has identified as alternative initiation sites.

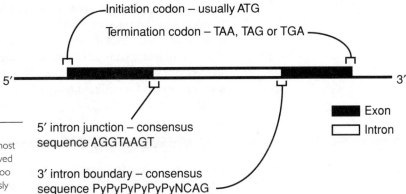

Initiation codon – usually ATG

Termination codon – TAA, TAG or TGA

5'

3'

■ Exon
□ Intron

5' intron junction – consensus
sequence AGGTAAGT

3' intron boundary – consensus
sequence PyPyPyPyPyPyNCAG

Fig 4.17

The structure of a eukaryotic gene containing a single intron. Although most exon-intron boundaries have conserved sequences, these are too short and too variable to be identified unambiguously

backwards, and you have no way of knowing which your gene lies in. This is not really a problem though as the computer enjoys this sort of task and is quite happy to work on it while you have a cup of coffee.

ORF searches are less useful if you are working with eukaryotic genomic DNA. This is because eukaryotic genes are frequently split into exons and introns. Exons within a gene may be very short (just a few codons) and are bounded not by initiation and termination codons, but by the exon-intron junctions (*Figure 4.17*). At present, we do not have efficient ways of recognizing these boundaries so cannot write a computer program to identify the protein coding regions of a split gene hidden in a eukaryotic DNA sequence. The ORF search can help you identify the longer exons, but you may not be able to locate the entire gene unless you know the amino acid sequence of the relevant protein or have some informative genetic data.

3.4 Comparing your sequence with other known sequences

Have you sequenced a completely novel piece of DNA or has someone else got there before you? To find the answer you must compare your sequence with all the published and unpublished sequences that have ever been obtained. This is not as impossible as it sounds as virtually all these DNA sequences are contained in the three international DNA Databases (see Section 3.5). By comparing your sequence with the databases you can determine if you do indeed have something new.

Although there are three Databases, they contain identical information and differ only in the countries they serve. If you wish you can obtain the entire current database on a CD-ROM and search it yourself, using your own personal computer (PC) and software. At present, a complete search takes about an hour with the standard types of PC used in research labs. This is long time to tie up a PC and there is a second disadvantage in that the latest CD-ROM release may not be entirely up to date. The databases grow by approximately 5 million nucleotides every 3 months, so if you are using the last quarter's CD-ROM then you are missing out on a lot of information. The alternative, and better, way to conduct the search is to send your sequence via electronic mail (e-mail) to the Database, who will conduct the search for you and relay the results back through the e-mail, usually within 24 hours.

3.4.1 Homology searches

Even though the databases currently contain over 100 million nucleotides there is still a good chance that your sequence is unique. If this is the case then you may be able to obtain valuable information by performing a homology search. This tells you if your sequence is similar to one or more sequences already in the database.

A homology search can be carried out with a DNA sequence but this is rarely of much value. The simplicity of DNA sequence information (just four different letters—A, C, G, and T) means that unrelated sequences can have quite high degrees of similarity, and identifying biologically significant homologies is very difficult. A homology search is more useful when carried out with an amino acid sequence translated from an ORF located in the DNA sequence. Amino acid sequence information is relatively complex (20 different letters), so two amino acid sequences with >50 per cent positional identity are likely to be biologically related (*Figure 4.18*). You may therefore be able to deduce the function of a gene within your DNA sequence by searching the database for homologous amino acid sequences.

◇ For full details on how these various analyses are carried out consult Cherry (1993).

There are several programs for homology searching, a popular one being FASTP (Pearson and Lipman 1988). This program can be run on a PC, using a CD-ROM database, or the service can be requested with the up-to-date database via the e-mail. A FASTP print-out lists homologous sequences with a numerical score for each one. The score expresses the degree of similarity that the computer has been able to find between the test sequence and the matching sequence. It is not a measure of biological relatedness and you still have to make your own judgement as to whether or not the similarity between the two sequences is important.

```
-GRKRKKRTSIETNIRLTLEKRFQDNPKPSSEEISMIAEQLSMEKEVVRVWFCNRRQKEKRIN-
 || |||||||||| |   ||| |   | || ||||  ||||| ||||| |||||||||||||||
-GRRRKKRTSIETNVRFALEKSFLANQKPTSEEILLIAEQLHMEKEVIRVWFCNRRQKEKRIN-
```

Fig 4.18

The alignment between two homologous amino acid sequences. The upper sequence is translated from a region within a gene expressed in rat epidermis. The alignment shows that this sequence is homologous with the POU homeodomain of the Oct-2 transcriptional regulator (lower sequence). The homology suggests that the epidermal gene product also possesses a POU homeodomain and so is also a transcriptional regulator. The amino acid sequences are shown in the one-letter code. (Taken from Anderson, B. *et al.* (1993) *Science*, 260, 78)

3.5 Submitting your sequence to a Database

When your sequence is complete you should deposit it in one of the Databases so other researchers can make use of the information. Many journals now insist that you do this before they publish the paper describing your results, and it is now recognized that sequence submission is a form of publication in its own right.

Table 4.2 The three DNA databases

Database	Sponsor
EMBL Data Library	European Molecular Biology Laboratory, Heidelberg, Germany
GenInfo/Backbone	National Center for Biotechnology Information, Bethesda, USA
DNA Data Bank of Japan	National Institute of Genetics, Mishima, Japan

◇ A typical accession number: X15011, the mitochondrial gene for NADH dehydrogenase subunit 5 of *Aspergillus nidulans*.

There are three DNA Databases (*Table 4.2*), and the one you use depends on where you are located. The three Databases exchange new sequences every day so your sequence quickly appears in all three. You can submit your sequence typed on a form, but it is better to send it through the e-mail. With 5 million nucleotides coming in every 3 months the Database people are quite busy so do not want to have to type your sequence into their computer by hand. Once your sequence has been received you will be sent an accession number which you or anyone else can use to retrieve a copy of the sequence in the future. Now at last you can put your feet up—you have completed your sequencing experiment.

Further reading

Cherry, J.M. (1993). Computer analysis of DNA and protein sequences. In *Current protocols in molecular biology*, (ed. F.M.Ausubel *et al.*), pp. 7.7.1–7.4.31. Greene Publishing Associates and John Wiley and Sons, New York—an excellent description of all aspects of computer-based sequence analysis.

Slatko, B.E. and Albright, L.M. (1993). Denaturing gel electrophoresis for sequencing. In *Current protocols in molecular biology*, (ed. F.M.Ausubel *et al.*), pp. 7.6.1–7.6.13. Greene Publishing Associates and John Wiley and Sons, New York—comprehensive protocols for sequencing gels.

Slatko, B.E., Albright, L.M., and Tabor, S. (1993). DNA sequencing by the dideoxy method. In *Current protocols in molecular biology*, (ed. F.M.Ausubel *et al.*), pp. 7.4.1–7.4.27. Greene Publishing Associates and John Wiley and Sons, New York—includes an excellent description how to identify problems from the appearance of the sequencing autoradiograph.

Ward, E.S. and Howe, C.J. (1989). Troubleshooting in chain-termination DNA sequencing. In *Nucleic acids sequencing: A practical approach* (ed. C.J.Howe and E.S.Ward), pp. 1–36. IRL Press at Oxford University Press—how to identify and solve problems that might arise during chain termination sequencing.

References

Andrews, A.T. (1991). Electrophoresis of nucleic acids. In *Essential molecular biology: A practical approach*, Vol. I (ed. T.A.Brown), pp. 89–126. IRL Press at Oxford University Press.

Biggin, M.D., Gibson, T.J., and Hong, G.F. (1983). Buffer gradient gels and ^{35}S label as an aid to rapid DNA sequence determination. *Proceedings of the National Academy of Sciences, USA*, **80**, 3963.

Brown, N.L. (1984). DNA sequencing. *Methods in Microbiology*, **17**, 259.

Mundy, C.R., Cunningham, M.W., and Read, C.A. (1991). Nucleic acid labelling and detection. In *Essential molecular biology: A practical approach*, Vol. II, (ed. T.A.Brown), pp. 57–109. IRL Press at Oxford University Press.

Olsson, A., Moks, T., Muhlen, J., and Gaal, A.B. (1984). Uniformly spaced banding patterns in DNA sequencing gels by the use of a field-strength gradient. *Journal of Biochemical and Biophysical Methodology*, **10**, 83.

Pearson, W.R. and Lipman, D.J. (1988). Improved tools for biological sequence comparison. *Proceedings of the National Academy of Sciences, USA*, **85**, 2444.

5 Chemical degradation sequencing

Those of you who have read this book assiduously page by page from the start of Chapter 1 will by now be thinking one of two things: either

- so that's how you do DNA sequencing, I wonder what the rest of the book is about? or
- I wonder when he will stop singing the praises of chain termination sequencing and get on to things that I'm interested in?

If you fall into the second category then thank you for your patience, we will now turn our attention to the second method for DNA sequencing—the chemical degradation technique first described by Allan Maxam and Walter Gilbert in 1977.

1. Chemical degradation sequencing in outline

If you have read the previous four chapters then you will have picked up quite a lot of information relevant to chemical degradation sequencing. The basic idea behind the two DNA sequencing methods is the same. In both cases the objective is to generate families of DNA molecules which, when separated by polyacrylamide gel electrophoresis, produce a banding pattern from which the sequence of the starting molecule can be read. The difference lies in the way in which the A, C, G, and T families of molecules are generated. In a chemical degradation experiment these families are produced not by synthesizing new strands (as occurs in chain termination sequencing) but by breaking down existing molecules with chemicals that cleave polynucleotides specifically at A, C, G, or T nucleotide positions.

To illustrate the way the technique works we will follow a brief outline of the steps that generate the G family of molecules in a chemical degradation experiment (*Figure 5.1*). The starting polynucleotides that we intend to sequence must have a radioactive marker at one end. We treat the molecules with a reagent—dimethyl sulphate—which reacts specifically with G nucleotides, resulting in a

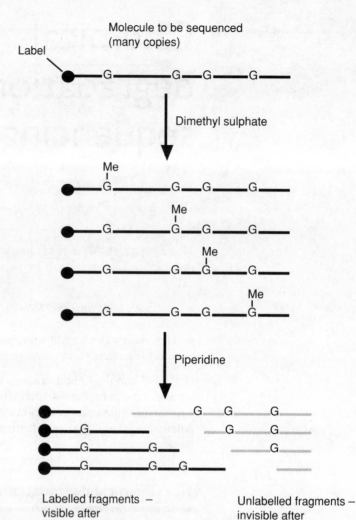

Fig 5.1

Generation of the G family of molecules in a chemical degradation sequencing experiment. Me = methyl group

Labelled fragments – visible after autoradiography

Unlabelled fragments – invisible after autoradiography

◇ Dimethyl sulphate methylates the N at position 7 of the purine ring. This in turn leads to a breakage of the bond between C8 and N9 of the ring, resulting in the modified structure that is recognized and removed by piperidine.

chemical modification to the purine ring. The amount of dimethyl sulphate that is added is carefully controlled so that on average we modify just one G nucleotide in each copy of the molecule to be sequenced. At this stage the DNA strands are still intact. Strand cleavage occurs when a second chemical—piperidine—is added to the reaction. Piperidine removes the modified G nucleotide and cuts the DNA molecule at the phosphodiester bond immediately up-stream of the 'base-less' site that is created.

Each cleavage produces two fragments, one from either side of the cut point. But only one of these fragments is labelled, so the second is in effect 'invisible' as it does not show up after autoradiography of the sequencing gel. Each cleavage therefore gives rise to a single band on the autoradiograph (*Figure 5.2*), the position of this band being determined by the length of the fragment, which in turn is set by the position within the original molecule of the G nucleotide that was modified during the chemical treatment. There were many copies of

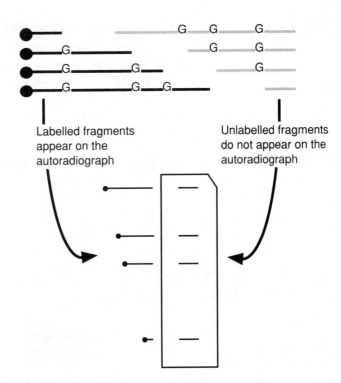

Labelled fragments
appear on the
autoradiograph

Unlabelled fragments
do not appear on the
autoradiograph

Fig 5.2

Only the labelled molecules give rise to a
band after autoradiography

our starting molecule, each of which has now been cleaved at a
phosphodiester bond adjacent to a G nucleotide. The G family of
fragments therefore produces a banding pattern that indicates the
positions of the Gs in the starting molecule. Comparison with the
banding patterns produced by the A, C, and T families enables the
sequence to be deduced in a manner similar to that used when reading the autoradiograph from a chain termination experiment (see
Figure 1.5).

1.1 The three steps in a chemical degradation sequencing experiment

The strand cleavage reactions comprise the central step in the
sequencing experiment:

PREPARE YOUR DNA

↓

CARRY OUT THE CHEMICAL
DEGRADATION REACTIONS

↓

RUN YOUR GEL

Step 1: Preparation of the end-labelled DNA
Step 2: The chemical degradation reactions
Step 3: Gel electrophoresis, autoradiography, and reading the
sequence

We will examine each step in turn and then, at the end of the chapter,
look at the relative merits of the chemical degradation and chain
termination sequencing procedures. This will help you understand
how to choose the most appropriate technique for your particular
sequencing project.

2. Preparing end-labelled DNA for a chemical degradation sequencing experiment

You can see from *Figures 5.1 and 5.2* that a key requirement with a chemical degradation experiment is that the DNA to be sequenced must be end-labelled. We must therefore spend a few minutes looking at ways of obtaining end-labelled DNA.

2.1 End-labelling

There are two possible ways of end-labelling a DNA molecule (Mundy *et al.* 1991):

- by the attachment of a labelled phosphate or nucleotide to the 5′ or 3′ end of a polynucleotide (*Figure 5.3A*)
- by filling in the 5′ overhangs created at the ends of a double stranded molecule by some restriction enzymes (*Figure 5.3B*)

A. Attaching a label to one end of a polynucleotide

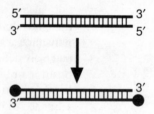

B. Filling in a 5′ overhang

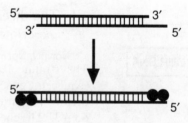

Fig 5.3

Two methods for end-labelling a double stranded DNA molecule

● Label

2.1.1 Attachment of a labelled phosphate or nucleotide

End-labelling is most frequently carried out with the enzyme called T4 polynucleotide kinase (PNK), which we met earlier in Chapter 3, Section 1.4.1. PNK transfers a phosphate group from the γ position of an ATP molecule onto the 5′ end of a dephosphorylated acceptor molecule. DNA molecules usually have phosphorylated 5′ ends, but an exchange reaction can be set up so that PNK removes these 5′ phosphates prior to carrying out the transfer. Alternatively, the phosphate groups can be removed by pretreating the DNA with alkaline phosphatase.

◇ The exchange reaction of PNK requires an excess of ADP in the reaction mixture.

Although PNK is the most popular enzyme for attaching radioactive markers to the ends of polynucleotides, it is not the only possibility. Terminal deoxynucleotidyl transferase (TdT) can also be used. TdT has two differences from PNK:

- the 3′ rather than 5′ end is labelled
- the label that is attached is a nucleotide, not a phosphate group

◇ Refer back to *Figure 5.2* and make sure you understand why the starting molecules must all be of the same length.

The first of these properties is not a problem, but the second can cause difficulties. TdT in fact adds a series of nucleotides to the 3′ end of each polynucleotide, and if left to its own devices would extend the strands by tens or even hundreds of nucleotides (*Figure 5.4*). If this happens then the polynucleotides are not all exactly the same length, which means they are not suitable for chemical degradation sequencing. The standard TdT reaction therefore has to be modified by replacing the labelled nucleotides with altered versions that cannot form long chains. In fact we use a labelled chain terminating nucleotide such as a ddNTP, or more usually cordycepin triphosphate. Cordycepin triphosphate, like the ddNTPs used in chain termination sequencing, lacks a hydroxyl group at the 3′ position on the sugar, and so once attached to a polynucleotide prevents further strand elongation. The result is that TdT adds only a single cordycepin triphosphate to each 3′ end (Tu and Cohen 1980).

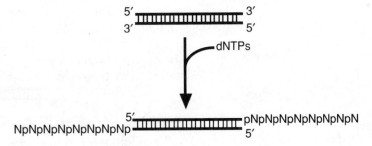

Fig 5.4

A possible result of treating a DNA molecule with terminal deoxynucleotidyl transferase

2.1.2 End-filling

Many restriction enzymes leave an 5′ overhang when they cut a double stranded DNA molecule. DNA polymerase enzymes are able to extend the recessed 3′ end, using the overhang as a template in a 'filling-in' reaction (see *Figure 5.3B*). If a labelled nucleotide is included in the reaction then the polynucleotides become 3′ end-labelled, presuming that the labelled nucleotide is complementary to one of the nucleotides in the overhang. Filling in reactions are usually carried out with Klenow polymerase, as an enzyme that lacks the 5′→3′ exonuclease activity is needed (see Chapter 3, Section 1.2.2). If

◇ The labelled nucleotide for end-filling must be chosen with care. You cannot, for instance, use labelled dCTP when filling in the overhangs produced by *Eco*RI as these overhangs have the sequence 3′-TTAA-5′.

this exonuclease activity is present (as in DNA polymerase I, for example) then the overhangs may be removed before the polymerase has a chance to fill them in.

2.2   Choosing a label

When we thought about labels for chain termination sequencing (Chapter 3, Section 1.4) we decided that ^{35}S is a better choice than ^{32}P, as ^{35}S gives sharper bands on the autoradiograph and is less hazardous. We appreciated that with ^{35}S the autoradiograph would have to be exposed for longer (as the emission energy of ^{35}S is less than that for ^{32}P), but we looked on this as a minor irritant.

With chemical degradation sequencing we have to consider an additional point, one that makes ^{35}S a much less suitable choice than ^{32}P. After a chemical degradation experiment the polynucleotides that are loaded on to the sequencing gel each carry just a single label. In contrast the polynucleotides from a chain termination experiment are multiply labelled, labelled nucleotides having been incorporated into the growing molecules during strand synthesis (see *Figure 3.11*). What this means is that the autoradiograph from a chemical degradation experiment must be exposed for longer than one from a chain termination gel. The weaker energy of ^{35}S now becomes a distinct disadvantage as it means that exposure for a week or more is needed before a clear autoradiograph is obtained. ^{32}P is therefore the more popular radioactive label for the chemical degradation method.

What about non-radioactive labels? In Chapter 3, Section 1.4.1 we looked at strategies based on chemiluminescent and fluorescent labels. With both of these there is sufficient sensitivity with a single end-label to detect the bands in the sequencing gel, and so both methods are also applicable to chemical degradation sequencing. However, these techniques are still new and their potential and limitations have not yet been fully assessed.

2.3   Converting the end-labelled molecule into a form suitable for sequencing

In Section 2.1 we learnt how to attach labels to the ends of linear, double stranded DNA molecules such as restriction fragments. However, these end-labelled, double stranded molecules cannot themselves be used in a chemical degradation sequencing experiment as they would give rise to two sequences, one for each polynucleotide. Only under the most exceptional circumstances will the two polynucleotides have the same sequence, so the eventual result will be two banding patterns superimposed on the autoradiograph. If this occurs then neither sequence will be readable.

A DNA molecule for chemical degradation sequencing must therefore be in one of two forms (*Figure 5.5*):

◇ Remember that complementary polynucleotides do not have the same sequence!

Single stranded, end-labelled

Double stranded, one
polynucleotide end-labelled

 Label

Fig 5.5

Suitable molecules for chemical
degradation sequencing

- single stranded, end-labelled
- double stranded, but comprising one end-labelled polynucleotide and one unlabelled polynucleotide

The second possibility, an asymmetrically labelled molecule, is acceptable as the unlabelled polynucleotide does not give rise to bands on the autoradiograph and so is invisible in the sequencing experiment.

2.3.1 Obtaining single stranded DNA

Of the two alternatives, obtaining single stranded DNA is the more difficult. We cannot use an M13 vector as the product of M13 cloning is circular single stranded DNA, and converting these circular molecules to linear polynucleotides, each molecule cut at the same position rather than randomly, would involve a series of quite complicated and difficult manipulations. In the long run this is simply not a viable approach.

The only possibility is to denature a double stranded DNA molecule so that the two polynucleotides become separated, and then to try to purify one or other of the strands. This can sometimes be achieved by denaturing polyacrylamide gel electrophoresis, as the two polynucleotides may have different purine:pyrimidine contents (*Figure 5.6*). If one strand has more purines than the other then it is slightly heavier and migrates more slowly in the gel. Two bands are therefore seen, one for each polynucleotide, and the desired strand can be purified by electroelution or some other means.

In practice this is not a good approach as it is not always possible to separate the polynucleotides in this way (sometimes the

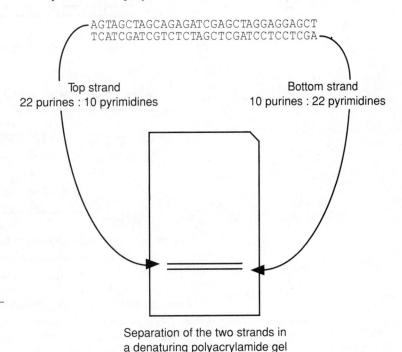

Fig 5.6

Denaturing polyacrylamide gel
electrophoresis can separate two
polynucleotides with differing
purine:pyrimidine contents

purine:pyrimidine contents are too close for the bands to be separated in the gel). Also, because the difference between the mobilities of the two polynucleotides is never great, the 'pure' polynucleotide that is obtained is frequently contaminated with a small amount of the second polynucleotide. This is often enough to give ghost bands that cause problems when the sequence is being read from the autoradiograph.

2.3.2 Obtaining asymmetrically labelled, double stranded DNA

As it is so difficult to obtain end-labelled single stranded DNA, the molecules used in chemical degradation sequencing experiments are usually double stranded but asymmetrically labelled. There are two ways of obtaining these molecules:

Restriction of a symmetrically labelled, double stranded molecule

If the double stranded molecule that has been end-labelled contains an internal restriction site, then it can be cut with the appropriate restriction enzyme and the two fragments separated in an agarose or polyacrylamide gel. The fragment required for sequencing, which is now asymmetrically labelled, can then be purified from the gel (*Figure 5.7*). This is a simple and effective approach, but in order to use it you must know what, if any, restriction sites are contained within the starting molecule. In many cases this information is not available until *after* the sequencing experiment.

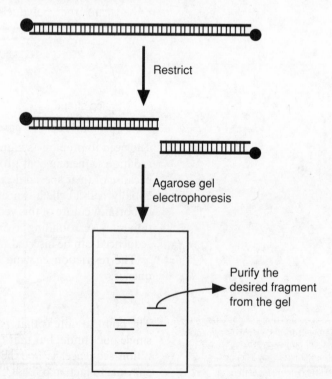

Fig 5.7

Using restriction to obtain asymmetrically
labelled, double stranded DNA

A. Inserting DNA into pSP64CS

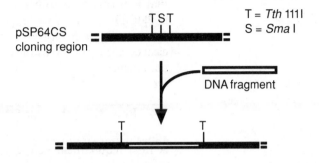

B. Recovery of the cloned fragment

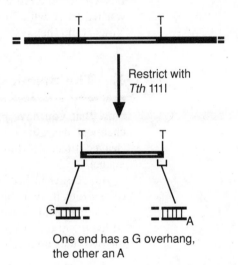

One end has a G overhang,
the other an A

Fig 5.8

Using pSP64CS

Cloning with a special type of vector

The need to prepare asymmetrically labelled DNA for chemical degradation sequencing has prompted the development of vectors which, because of their special design, enable cloned fragments to be differentially end-labelled. An example is pSP64CS (Eckert 1987). The important feature of this vector is the presence of two *Tth*111I restriction sites surrounding a *Sma*I site into which the fragment to be sequenced can be inserted (*Figure 5.8A*).

The restriction enzyme *Tth*111I has an unusual recognition sequence:

$$5'\text{-GACN}\!\downarrow\!\text{NNGTC-}3'$$

◇ Of course, complications arise with these vectors if the cloned fragment has one or more internal *Tth*1 1 1I sites, but this is not highly likely as the enzyme is a relatively infrequent cutter.

The point to note is that *Tth*111I leaves a 5′ overhang comprising a single nucleotide, but that this nucleotide, being an 'N' in the recognition sequence, is not the same at every terminus created by the enzyme. In fact, in pSP64CS the sequence is such that after restriction

◇ A second advantage of pSP64CS and related vectors is they also enable progressive deletions to be made in the cloned fragment, in a manner similar to that described in Chapter 2, Section 1.4.3.

with *Tth*111I the insert DNA is recovered as a fragment with a G overhang at one end, and an A at the other (*Figure 5.8B*). Now all that is necessary is to carry out an end-filling reaction with either [^{32}P]dCTP or [^{32}P]dTTP to specifically label one end of the fragment. Note that you do not even need to purify the fragment from the cleaved vector DNA, as the latter has C and T overhangs, which are not labelled if the appropriate nucleotide is chosen.

2.4 Removing the salt from the DNA preparation

A final requirement before you start sequencing is that the DNA preparation must be free of salt. If salt is present then the chemical degradation reaction for T nucleotides is suppressed, resulting in faint bands in the T lane of the autoradiograph. After labelling, the DNA must therefore be precipitated with ethanol, and the pellet washed twice with 70 per cent ethanol to remove all traces of salt. You should then resuspend the DNA in a suitable volume of water.

3. The chemical degradation reactions

Now that you have prepared your DNA you can embark on the chemical degradation reactions that will provide the families of cleaved molecules that you will subsequently load on to the sequencing gel.

As we saw in Section 1, the chemical degradation reactions are carried out in two steps (see *Figure 5.1*). First, the DNA is treated with a chemical that modifies a nucleotide base. Then piperidine is added to cut the DNA at the modification position. What exactly do the two steps involve?

3.1 The nucleotide modification reactions

◇ When deciding how much DNA to use the important factor is the amount of labelling that has been achieved. To obtain a clear autoradiograph your starting DNA must be labelled to a specific activity of at least 5×10^4 d.p.m. μg^{-1}. To achieve this you will need about 0.5 μg of a restriction fragment (>1000 bp) or 100 ng of a synthetic oligonucleotide (<100 nucleotides).

If you are following the standard sequencing procedure, you will divide your DNA into four aliquots and treat the tubes as follows (*Table 5.1*):

- tube 1—add dimethyl sulphate (the 'G' reaction)
- tube 2—add formic acid (the 'A+G' reaction)
- tube 3—add hydrazine (the 'C+T' reaction)
- tube 4—add NaCl plus hydrazine (the 'C' reaction)

Two of the reactions are not entirely base-specific. It has proved impossible to devise modification reactions that recognize each base individually, and in this standard procedure two of the reactions give mixed families of molecules, an 'A+G' family and a 'C+T' family. As we will see in Section 4, this does not usually cause any problems when we come to read the sequencing gel. If you wish, you can carry out a fifth reaction (with NaOH plus EDTA—see

◇ *CAUTION* These chemicals are chosen because they react with DNA. If given half a chance they will happily modify *your* DNA as well as the DNA you wish to sequence. Both dimethyl sulphate and hydrazine are highly toxic by inhalation or contact with skin. All the chemicals should be handled in a fume hood, taking the necessary precautions, and disposed of in the appropriate way. If you have any doubts talk to your safety representative.

Table 5.1 Nucleotide modification reactions used in chemical degradation sequencing

Reaction	Reagents
G	Dimethyl sulphate
A+G	Formic acid
C+T	Hydrazine
C	NaCl + hydrazine
A > C	NaOH + EDTA
Modified A > C	NaOH
T	Potassium permanganate

Table 5.1) that cleaves at A and C nucleotides, with a preference for As. This 'A > C' reaction provides you with a fifth lane on the sequencing gel and helps to clear up any confusion that might arise. Most of the time it is not needed.

The standard procedure is the most popular version of the chemical degradation method. Variations have been developed over the years (e.g. Ambrose and Pless 1987), one objective being to replace the more hazardous chemicals with less dangerous ones, but the standard procedure has remained the most widely used. A couple of the alternative reactions are described in *Table 5.1*.

Each modification is complete after a few minutes at 25°C. You must then stop the reactions by adding an acetate buffer (the exact formulation depending on the reaction), and recover the DNA by ethanol precipitation. You are now ready to move on to the second stage of the procedure.

3.2 Strand cleavage

◇ *CAUTION* Piperidine is highly flammable and toxic by inhalation and contact with skin. Note that stock solutions of piperidine (usually 10 M) dissolve most plastics and so should not be stored in plastic bottles.

To carry out the strand cleavages you resuspend the DNA pellets from the ethanol precipitation in 1 M piperidine and incubate for 30 minutes at 90°C. During this incubation the modified purine and pyrimidine bases are removed from the polynucleotides, and each strand is cut at the phosphodiester bond immediately upstream of (i.e. to the 5′ side of) the base-less position.

The key necessity at this stage is to make sure that the cleavage reactions go to completion. If many modified sites remain uncut there will be a preponderance of full length polynucleotides in the reaction mixtures. This in turn means fewer cleaved molecules and as a result the bands on the autoradiograph will be weak and indistinct. Thirty minutes at 90°C is sufficient time for the cutting reactions to be completed, but you must make sure the tubes are sealed tightly so that piperidine is not lost by evaporation. Loss of piperidine may result in an incomplete reaction. After incubation the piperidine is removed in a rotary evaporator and the DNA samples resuspended in loading buffer, ready for electrophoresis in the sequencing gel.

4. Running the gel and reading the sequence

The sequencing gel is prepared, electrophoresed, and autoradiographed exactly as described for the chain termination method. You should therefore read Chapter 4, Section 1, if you have not already done so.

The differences arise when you come to read the sequence from the autoradiograph (*Figure 5.9*). Remember that two of the modification reactions are specific not for one but for two different nucleotides. As a result there will be some bands on the autoradiograph that span two tracks rather than just one. However, as you can see from *Figure 5.9*, an unambiguous sequence can still be read.

In fact, unambiguity is one of the main advantages of the chemical degradation method. In Chapter 4, Section 2, we spent a good bit of time dealing with the various problems that can make it difficult to interpret the banding pattern on an autoradiograph after a chain termination sequencing experiment. The chain termination method, requiring strand synthesis by enzymatic activity, is inherently more complex than simple chemical degradation. Most of the problems that arise during the chain termination method are due to the strand synthesis reactions not working properly (e.g. inappropriate dNTP:ddNTP ratios, a poor quality enzyme, hairpin loops in the template). These problems do not occur in the chemical degradation method, and as a result the autoradiograph is usually much less

◇ For an excellent description of problem solving, see Maxam and Gilbert (1980).

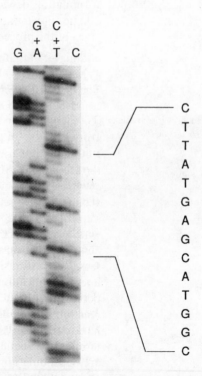

Fig 5.9

The autoradiograph from a chemical degradation sequencing experiment. (Reproduced with permission from Eckert (1992), page 7.5.6 ©1992).

trouble to read. Those problems that might arise are generally easy to correct:

- the sequence in one or more lanes may be faint at the top or bottom of the autoradiograph, indicating that the modification reactions were allowed to proceed for too long or not long enough respectively
- bands representing the Ts are weak, because salt was not completely removed from the DNA preparation
- there are smears in the lanes, due to the piperidine not being completely removed

Finally, as with chain termination sequencing, the presence of hairpin loops in the cleaved molecules may lead to band compressions in one or more lanes (see Chapter 4, Section 2.3.1). The solution here is to add an extra denaturant (e.g. formamide) to the gel, to electrophorese the gel at a higher voltage, or to sequence the other strand of the starting DNA molecule.

5. Choosing the best sequencing method for your own project

In today's molecular biology labs the chain termination method is by far the most popular and frequently used procedure for DNA sequencing. But this has not always been the case. The two methods were developed almost simultaneously in the mid-1970s, and at first most labs used the chemical degradation method. This was primarily because chain termination sequencing requires special reagents (modified polymerases, oligonucleotide primers) that were not easily obtainable at that time, and also involves specialized techniques such as M13 cloning, which not many research groups were familiar with. The chemical degradation method was much easier to carry out: it could be applied directly to double stranded restriction fragments and made use of standard organic chemicals that were readily available.

Gradually during the 1980s the chain termination method gained the ascendancy. Many suppliers starting manufacturing good quality enzymes and primers, and the development of the M13mp series of vectors made it relatively easy to obtain single stranded DNA. Subsequent improvements in the strand synthesis reactions, together with the introduction of specialized sequencing enzymes such as Sequenase, mean that nowadays it is possible to obtain 500 bp of sequence from a single chain termination experiment. Even with the latest modifications the upper limit for a chemical degradation experiment is about 250 bp. In addition, the modification and cleavage reactions take longer to carry out, involve more steps, and are more hazardous than the strand synthesis reactions required for chain termination sequencing. Not surprisingly, the big genome projects are depending almost exclusively on chain termination sequencing, and most labs use this method in their day-to-day experiments.

The temptation is to downgrade the chemical degradation method. If you read a book on DNA sequencing, even a very reputable one, you will probably find that the author spends most of the time describing chain termination sequencing, relegating the chemical degradation method to a single chapter at the end of the book. But chemical degradation sequencing has a number of distinct advantages that makes it, for some applications at least, the superior method. For example:

- Chemical degradation sequencing is less prone to problems caused by 'unusual' structures in the DNA being sequenced. Hairpin loops are not such a big problem, and long runs of the same nucleotide (e.g. CCCCCCCCCCCCCCCCCC) can be sequenced without difficulty. DNA polymerases tend to get

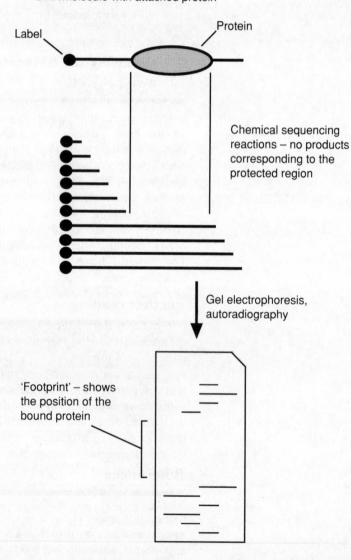

DNA molecule with attached protein

Label

Protein

Chemical sequencing reactions – no products corresponding to the protected region

Gel electrophoresis, autoradiography

'Footprint' – shows the position of the bound protein

Fig 5.10

Footprinting

confused in these homopolymer runs and occasionally skip one or more nucleotides. This results in deletions in the chain terminated molecules, leading to shadow bands on the autoradiograph. If you have used homopolymer tailing (Deng and Wu 1983) at some stage in your cloning procedure, then you may find it very difficult with the chain termination method to sequence through the tails into your inserted DNA fragment. Similarly, cDNA clones with long poly(A) tails (derived from polyadenylated mRNA) may cause problems.

- If you wish to check the sequence of a synthetic oligonucleotide, then you should use the chemical degradation method. Cloning the oligonucleotide prior to chain termination sequencing is a lengthy and tedious business. It is much easier to end-label the oligonucleotide and sequence it directly by the chemical degradation method.

- The chemical degradation method enables you to study interactions between proteins and DNA molecules. If the DNA molecule you are sequencing has a protein attached (e.g. at an enhancer or other control site) then some of the nucleotides are 'protected', meaning that they are inaccessible to the modification chemicals (*Figure 5.10*). The result is a 'footprint' on the autoradiograph, a region where no bands occur (Galas and Schmitz 1978). The position of this footprint pinpoints the protein attachment site on the DNA molecule.

You can see from this list that there are specific applications for which the chemical degradation method is well suited. If your project involves this sort of work then you will need to use the procedure. However, if the nature of your work means that there is no special advantage in using chemical degradation sequencing the advice is to stick with the chain termination method. The latter is quicker to set up, less hazardous, and has the potential to generate your desired sequence with the minimum number of experiments.

Further reading

Eckert, R.L. (1993). DNA sequencing by the chemical method. In *Current protocols in molecular biology*, (ed. F.M.Ausubel *et al.*), pp. 7.5.1–7.5.11. Greene Publishing Associates and John Wiley and Sons, New York—this is currently the clearest description of the chemical degradation sequencing procedures.

Barker, R.F. (1989). Maxam and Gilbert sequencing using one metre gel systems. In *Nucleic acids sequencing: A practical approach*, (ed. C.J.Howe and E.S.Ward), pp. 117–135. IRL Press at Oxford University Press—useful for its description of special gels to maximize the sequence information obtainable by this method.

References

Ambrose, B.J.B. and Pless, R.C. (1987). DNA sequencing: chemical methods. *Methods in Enzymology*, **152**, 522.

Deng, G.R. and Wu, R. (1983). Terminal transferase: use in the tailing of DNA and for in vitro mutagenesis. *Methods in Enzymology*, **100**, 96.

Eckert, R. (1987). New vectors for rapid sequencing of DNA fragments by chemical degradation. *Gene*, **51**, 245.

Galas, D. and Schmitz, A. (1978). DNase footprinting: a simple method for the detection of protein-DNA binding specificity. *Nucleic Acids Research*, **5**, 3157.

Maxam, A.M. and Gilbert, W. (1977). A new method for sequencing DNA. *Proceedings of the National Academy of Sciences, USA*, **74**, 560.

Maxam, A. and Gilbert, W. (1980). Sequencing end-labeled DNA with base-specific chemical cleavages. *Methods in Enzymology*, **65**, 499.

Mundy, C.R., Cunningham, M.W., and Read, C.A. (1991). Nucleic acid labelling and detection. In *Essential molecular biology: A practical approach*, Vol. II, (ed. T.A. Brown), pp. 57–109. IRL Press at Oxford University Press.

Tu, C.-P.D. and Cohen, S.N. (1980). 3′-end labeling of DNA with [α-^{32}P]cordycepin-5′-triphosphate. *Gene*, **10**, 177.

Glossary

Accession number An alphanumeric code provided by a Database as an identifier for a individual DNA sequence.

Acrylamide One of the monomeric chemical components of a polyacrylamide gel.

Alkaline lysis A procedure for the preparation of double stranded plasmid DNA from *E. coli* cells.

Ammonium persulphate The catalyst in polymerization of an acrylamide gel.

Autoradiography A method of detecting radioactively labelled molecules through exposure of an X-ray sensitive photographic film.

Bacteriophage or Phage A virus whose host is a bacterium. Bacteriophage DNA molecules are often used as cloning vectors.

Biotin A molecule that can be incorporated into dUTP and used a non-radioactive label for a DNA molecule.

Bis See *N,N′*-methylenebisacrylamide.

Bromophenol blue A marker dye used in polyacrylamide gel electrophoresis.

Buffer gradient gel A gel in which the buffer concentration is greater at the bottom than at the top, designed to give more even spacing between bands.

Caesium chloride density gradient centrifugation A procedure for separating DNA molecules of different buoyant densities by centrifugation in a concentrated solution of caesium chloride; often used to purify plasmid DNA from bacterial DNA.

Chemiluminescence The chemical production of light, used in non-radioactive labelling procedures.

Compression A region of a sequencing autoradiograph in which the bands have become compressed together due usually to formation of hairpin loops in the polynucleotides being electrophoresed.

Contig A contiguous (i.e. continuous) DNA sequence built up from a series of shorter overlapping sequences.

Cordycepin triphosphate A modified nucleotide sometimes used in end-labelling reactions with terminal deoxynucleotidyl transferase.

Database   An international depository for nucleic acid and protein sequence information.

7-deaza-dGTP   A modified nucleotide sometimes used to reduce the formation of hairpin loops in chain terminated polynucleotides.

Deoxyinosine triphosphate (dITP)   A modified nucleotide sometimes used to reduce the formation of hairpin loops in chain terminated polynucleotides.

Deoxyribonuclease I   A non-specific endonuclease often used to make random breaks in a DNA molecule.

Dideoxynucleotide   A modified nucleotide that lacks the 3' hydroxyl group and so prevents further strand synthesis when incorporated into a growing polynucleotide.

Digitizer   An electronic analyser that enables DNA sequence information to be entered into a computer directly from the autoradiograph.

Dimethyl sulphate (DMS)   The modification chemical used in the G reaction of the chemical degradation sequencing procedure.

DNA polymerase   An enzyme that synthesizes DNA by polymerization of nucleotide subunits, usually in a template-dependent manner.

Electrophoresis   Separation of molecules on the basis of their net electric charge.

End-filling   Conversion of a sticky end to a blunt end by enzymatic synthesis of the complement to the single stranded overhang.

End-labelling   Attachment of a marker to the end of a polynucleotide.

Ethanol precipitation   Precipitation of nucleic acid molecules by ethanol plus salt, used primarily as a means of concentrating DNA.

Exonuclease III   An exonuclease that digests double stranded DNA from termini with 5' overhangs, used to make progressive deletions in a cloned DNA molecule.

FASTP   A popular computer program for homology searching.

Fixing   Post-treatment of a denaturing polyacrylamide gel by soaking in acetic acid to leach out the urea.

Fluorescence   A detection system used in a non-radioactive version of DNA sequencing.

Formamide   A chemical denaturant used in the loading buffer for a denaturing polyacrylamide gel, and sometimes included in the gel itself to reduce formation of hairpin loops in the polynucleotides being electrophoresed.

Formamide dye mix   A denaturing loading buffer used in the electrophoresis of DNA sequencing products.

Formic acid   The modification chemical used in the A+G reaction of the chemical degradation sequencing procedure.

Gel electrophoresis   Electrophoresis performed in a gel matrix so that molecules of similar electric charge can be separated on the basis of size.

Ghost bands   Extra bands that appear at random on a sequencing autoradiograph.

Helper phage   A phage that is introduced into a host cell in conjunction with a related cloning vector, in order to provide enzymes and other proteins required for replication of the cloning vector.

Homology search   A computer based search of a database to determine if a new sequence has sequence similarity with an existing sequence. The term 'homology' refers to evolutionary relatedness and should be used with care.

Hydrazine   The modification chemical used in the C and C+T reactions of the chemical degradation sequencing procedure.

IPTG   Isopropyl-thiogalactoside—a non-metabolizable inducer of the *lac* operon used in Lac selection.

Klenow polymerase   A DNA polymerase enzyme, obtained by modification of *E. coli* DNA polymerase I, used in chain termination sequencing.

Labelling-termination procedure   A version of chain termination sequencing in which the DNA polymerase is initially provided with limited quantities of dNTPs so that the newly synthesized strands become heavily labelled before subsequent extension and termination.

Lac selection   A means of identifying recombinant bacteria containing vectors that carry the *lacZ'* gene; the bacteria are plated on a medium that contains an analogue of lactose that gives a blue colour in the presence of β-galactosidase activity.

M13   A bacteriophage that infects *E. coli*, derivatives of which are used as cloning vectors for the preparation of single stranded DNA.

N,N'-methylenebisacrylamide (bis)   One of the monomeric chemical components of a polyacrylamide gel.

N,N,N',N'-tetramethylethylenediamine (TEMED)   The initiator in polymerization of an acrylamide gel.

Phagemid   A hybrid phage-plasmid vector used for synthesis of single stranded DNA.

Phenol extraction   A method used to remove protein from DNA preparations.

Pile-up   A series of strong bands in all four lanes of a chain termination autoradiograph, usually caused by a hairpin loop in the template DNA.

Piperidine The chemical used in the chemical degradation sequencing procedure to cleave polynucleotides at modified sites.

Plaque A zone of clearing on a lawn of bacteria caused by lysis of the cells by infecting phage particles.

Polyethylene glycol (PEG) A polymeric compound used to precipitate phage particles prior to DNA preparation.

Polylinker A synthetic double stranded oligonucleotide carrying a number of restriction sites.

Polymerase chain reaction (PCR) An enzymatic method for amplification of a region of a DNA molecule.

Primer A short single stranded oligonucleotide which, when attached by base pairing to a single stranded template molecule, acts as the start point for complementary strand synthesis directed by a DNA polymerase enzyme.

Processivity Refers to the length of new strand that a DNA polymerase is able to synthesize before dissociating from the template.

Progressive deletions A series of increasingly lengthy deletions made in a cloned DNA molecule as a means of obtaining subclones from which a long overlapping sequence can be built up.

Replicative form (RF) The double stranded form of the M13 DNA molecule found within infected *E. coli* cells.

Restriction enzyme An endonuclease that cuts double stranded DNA at specific recognition sequences.

Sequenase A modified version of T7 DNA polymerase, with high processivity and low or zero exonuclease activity, popular in chain termination sequencing.

Sex pilus A structure present on the surface of a bacterium containing a conjugative plasmid, used as the entry point for DNA from infecting M13 phage particles.

Shadow bands One or more additional bands immediately below an authentic band on a sequencing autoradiograph, usually caused by nucleotide deletions in the molecules that were electrophoresed.

Sharkstooth comb A special type of comb for loading a sequencing gel, designed so that the lanes are touching to make it easier to read the autoradiograph.

Silanizing solution 2 per cent dimethyldichlorosilane in 1,1,1-trichloroethane—a silicone polish used to prevent a polyacrylamide gel adhering to the gel plates when the assembly is taken apart after electrophoresis.

Sonication Exposure of DNA to ultrasonic vibrations in order to create random breaks.

Stem-loop A hairpin structure that may form in a polynucleotide.

***Taq* DNA polymerase**   A thermostable DNA polymerase used in PCR and some chain termination sequencing procedures.

TEMED   See *N,N,N′,N′*-tetramethylethylenediamine.

Template   A single stranded polynucleotide (or region of a polynucleotide) used to direct synthesis of a complementary polynucleotide.

Terminal deoxynucleotidyl transferase (TdT)   An enzyme used to attach radioactive nucleotides to the 3′ ends of DNA molecules.

Termination-chase procedure   A version of chain termination sequencing where ddNTPs are added at the outset of the strand synthesis reactions, and unterminated molecules are extended to full length by a subsequent chase reaction.

T4 polynucleotide kinase (PNK)   An enzyme used to attach radioactive phosphate groups to the 5′ ends of DNA molecules.

Thermal cycle sequencing   A version of chain termination sequencing that enables the sequence of a PCR product to be determined without prior cloning.

Transfection   Uptake of phage DNA by competent bacterial cells.

Universal primer   A primer for chain termination sequencing designed to anneal to all M13 based cloning vectors and so applicable to most DNA sequencing experiments.

Urea   A chemical denaturant used in polyacrylamide gel electrophoresis.

Wall   A series of strong bands in all four lanes of a chain termination autoradiograph, usually caused by a hairpin loop in the template DNA.

Wedge gel   A gel that is thicker at the bottom than at the top, designed to give more even spacing between bands.

X-gal   5-bromo-4-chloro-3-indolyl-β-D-galactopyranoside—a lactose analogue used as a colorimetric substrate in Lac selection.

Xylene cyanol   A marker dye used in polyacrylamide gel electrophoresis.

Index